T0139133

Emerging Green Technologies

Emerging Green Technologies

MATTHEW N. O. SADIKU

CRC Press
Taylor & Francis Group
Boca Raton London New York

CRC Press is an imprint of the
Taylor & Francis Group, an **informa** business

CRC Press
Taylor & Francis Group
6000 Broken Sound Parkway NW, Suite 300
Boca Raton, FL 33487-2742

First issued in paperback 2022

© 2020 Taylor & Francis Group, LLC
CRC Press is an imprint of Taylor & Francis Group, an Informa business

No claim to original U.S. Government works

ISBN-13: 978-0-367-36161-7 (hbk)
ISBN-13: 978-1-03-233677-0 (pbk)
DOI: 10.1201/9780429344213

Reasonable efforts have been made to publish reliable data and information, but the author and publisher cannot assume responsibility for the validity of all materials or the consequences of their use. The authors and publishers have attempted to trace the copyright holders of all material reproduced in this publication and apologize to copyright holders if permission to publish in this form has not been obtained. If any copyright material has not been acknowledged please write and let us know so we may rectify in any future reprint.

Except as permitted under U.S. Copyright Law, no part of this book may be reprinted, reproduced, transmitted, or utilized in any form by any electronic, mechanical, or other means, now known or hereafter invented, including photocopying, microfilming, and recording, or in any information storage or retrieval system, without written permission from the publishers.

For permission to photocopy or use material electronically from this work, access www.copyright.com or contact the Copyright Clearance Center, Inc. (CCC), 222 Rosewood Drive, Danvers, MA 01923, 978-750-8400. For works that are not available on CCC please contact mpkbookspermissions@tandf.co.uk

Trademark notice: Product or corporate names may be trademarks or registered trademarks, and are used only for identification and explanation without intent to infringe.

Publisher's Note

The publisher has gone to great lengths to ensure the quality of this reprint but points out that some imperfections in the original copies may be apparent.

Visit the Taylor & Francis Web site at
http://www.taylorandfrancis.com

and the CRC Press Web site at
http://www.crcpress.com

Contents

Preface

The three major challenges that mankind is facing today are population explosion, environmental degradation, and scarcity of resources. The technologies that build products and systems to help in confronting these challenges are known as green technologies. Technology refers to the application of knowledge for human benefit. Although green describes a color, in the 21st century, green has evolved to have a deeper meaning related to environmental issues. The 21st century has been regarded as an environment protection century.

Green technology deals with using science and technology to protect the environment as well as curb the negative impacts of human involvement. It is any mode of technology that lowers CO_2 emissions. Some people refer to green technology as sustainable technology, environmental technology, or clean technology. Here, green technology means efforts to promote sustainability and reduce greenhouse gas emissions. Green technology has been developed in response to challenges that humankind faces. The top ten problems/challenges identified by Nobel Laureate Richard E Smalley are as follows: (1) energy, (2) water, (3) food, (4) environment, (5) poverty, (6) terrorism and war, (7) disease, (8) education, (9) democracy, and (10) population.

Green technologies include green energy, green chemistry, green engineering, green IT, green food, green manufacturing, green business, green economics, green supply chain, green logistics, green building, and green nanotechnology. The author of this book has identified 15 key emerging green technologies, which will propel our economy in the near future. Their development will lead to global and sustainable powers that will impact our economics, societies, cultures, and our way of life. Several business establishments have used at least one green technology or practice in order to make their production processes more environmentally friendly.

The book is organized into 16 chapters. The first chapter is an introduction to green technologies. Chapter 2 is on green or renewable energy, which is a form of energy that does not contribute to climate change or global warming. It uses energy sources that are continually replenished by nature. Chapter 3 is about green chemistry and green engineering, which use scientific knowledge to reconcile the need for chemical production and the desire to reduce or eliminate the use of hazardous substances. Chapter 4 deals with the Green Revolution, which has changed the agriculture sector. Green Revolution refers to the transformation in agricultural practices in many developing nations. Chapter 5 dwells on green agriculture and foods, which are foods produced under the principle of sustainable development. Chapter 6 addresses green nanotechnology, which involves making green nanoproducts that are more environmentally friendly throughout their life cycle, and using them to support sustainability.

Chapter 7 deals with green computing, or green IT, which refers to the practice of reducing the environmental footprint of technology by efficiently using computing assets. Chapter 8 is on green Internet of things (IoT), which focuses on reducing the energy consumption of IoT, thereby fulfilling the smart world with sustainability. Chapter 9 is on green cloud computing, which seeks to achieve sustainable development of cloud computing and reduce the possible impact of cloud systems on the environment. Chapter 10 discusses green communications and networking that can introduce significant reductions in energy consumption in information and communication technology (ICT) systems. Chapter 11 focuses on green growth and green economy, which is an economy that aims at reducing environmental risks and ecological scarcities. It offers a potential solution to ecological sustainability, social justice, and lasting prosperity. Chapter 12 is on green business (or sustainable business), which refers to meeting customer's needs without causing environmental and social problems. It is making business operations more environmentally friendly (or "green"), since a green environment is a social as well as a business issue. In Chapter 13, we cover green marketing, which involves integrating concerns about the environment into the practice and principles of marketing. It is done by marketers to publicize and sell their green products to consumers. Chapter 14 is on green manufacturing, which refers to modern manufacturing that makes products without pollution. It is the embodiment of the strategy for sustainable development of the manufacturing sector. Chapter 15 is about green supply chain management, which is an emerging concept motivated by the need for environmental consciousness. It aims at integrating environmental thinking into supply chain. Finally, Chapter 16 addresses green logistics and green transportation, which basically consist of all activities related to the eco-efficient management of the forward and reverse flows of products between the point of origin and the point of consumption.

This book provides researchers, students, and professionals a comprehensive introduction to applications, benefits, and challenges of each green technology. It presents the impact of these cutting-edge technologies on our global economy and its future. The author was motivated to write this book partly due to the lack of a single source of reference on these technologies. There are monographs on each technology, but there is none that combines these technologies. Hence, the book will help a beginner in acquiring introductory knowledge about these emerging technologies. The main objective of the author is to provide a concise treatment that is easily digestible. It is hoped that the book will be useful to practicing engineers, computer scientists, and information business managers.

This monograph provides an introduction to popular emerging green technologies. It is a must-read for those graduate students or scholars who consider researching green technologies. It can also serve as a valuable resource for business professionals who seek ways to green their processes.

I am grateful to Dr. Pamela Obiomon, dean of the College of Engineering at Prairie View A&M University, and Dr. Kelvin Kirby, head of the Department of Electrical and Computer Engineering, for their constant support and appreciation. I would like to thank my students, Dr. Mahamadou Temberly, Dr. Emmanual Shadare, and Adedamola Omotoso, for our joint publications used in this book. I also want to thank Toyin Oso and Segun Odusote for helping with the quotations at the beginning of each chapter. Well-deserved gratitude goes to my wife Kikelomo Esther for her constant support and prayers.

Matthew N. O. Sadiku
Prairie View, Texas

About the Author

Matthew N. O. Sadiku received his B. Sc. degree in 1978 from Ahmadu Bello University, Zaria, Nigeria, and his M.Sc. and Ph.D. degrees from Tennessee Technological University, Cookeville, TN, in 1982 and 1984, respectively. From 1984 to 1988, he was assistant professor at the Florida Atlantic University, Boca Raton, FL, where he did graduate work in computer science. From 1988 to 2000, he was at Temple University, Philadelphia, PA, where he was made full professor. From 2000 to 2002, he was with Lucent/Avaya, Holmdel, NJ, as a system engineer and with Boeing Satellite Systems, Los Angeles, CA, as a senior scientist. He is presently professor of electrical and computer engineering at Prairie View A&M University, Prairie View, TX.

Dr. Sadiku has authored over 740 professional papers and over 80 books, including *Elements of Electromagnetics* (Oxford University Press, 7th ed., 2018), *Fundamentals of Electric Circuits* (McGraw-Hill, 7th ed., 2020, with C. Alexander), *Computational Electromagnetics with MATLAB* (CRC Press, 4th ed., 2019), and *Principles of Modern Communication Systems* (Cambridge University Press, 2017, with S. O. Agbo). In addition to the engineering books, he has authored books on Christianity, including *Secrets of Successful Marriages, How to Discover God's Will for Your Life*, and commentaries on all the books of the New Testament Bible. Some of his books have been translated into French, Korean, Chinese (and Chinese Long Form in Taiwan), Italian, Portuguese, and Spanish.

Dr. Sadiku was the recipient of the 2000 McGraw-Hill/Jacob Millman Award for his outstanding contributions in the field of electrical engineering. He also received the Regents Professor award for 2012–2013 from the Texas A&M University System. He is a registered professional engineer and a fellow of the Institute of Electrical and Electronics Engineers (IEEE) "for contributions to computational electromagnetics and engineering education." He was the IEEE Region 2 Student Activities Committee chairman and associate editor for IEEE Transactions on Education. Dr. Sadiku is also a member of the Association for Computing Machinery (ACM) and the American Society of Engineering Education (ASEE). His current research interests are in the areas of computational electromagnetics, computer networks, and engineering education. His works can be found in his autobiography, *My Life and Work* (Trafford Publishing, 2017) or on his website: www.matthew-sadiku.com. He currently resides with his wife Kikelomo in Hockley, Texas and can be reached via email at sadiku@ieee.org

1

Introduction to Green Technologies

> Engineering is the art of modelling materials we do not wholly understand, into shapes we cannot precisely analyze so as to withstand forces we cannot properly assess, in such a way that the public has no reason to suspect the extent of our ignorance.
>
> —A. Dykes

1.1 Introduction

Awareness about the impact of mankind's modern lifestyle on the environment has been rapidly increasing in recent years. This impact arises from pollution, consumption, and destruction of natural resources. This has led to greater awareness of the need for sustainable and environment-friendly practices. This awareness has led to the emergence of green technologies in recent years.

Technology refers to the application of knowledge for practical purposes and for human benefits. Green technology is one that takes into account the impact an invention has on the environment. Some people refer to green technology as sustainable technology, environmental technology, or clean technology. Here green technology is used to mean effort to promote sustainability and reduce greenhouse gas emissions. Green technologies and practices are those that lessen the impact of business operations on the environment. As with any technology, the development of green technology requires investment and initiative to support development projects. It also requires encouraging the whole society to participate in the green technology innovation and forming green consumption consciousness. Green technologies include green energy, green IT, green food, green manufacturing, green business, green economics, green supply chain, green logistics, green building, and green nanotechnology [1].

This chapter provides a brief introduction to green technology. It begins with explaining the concept of green technology. Then it encourages going green. It provides specific examples of green technologies, which will be fully discussed in the book. The advantages and disadvantages of green technologies are presented in comparison to conventional and more polluting technologies. The benefits and challenges of green technologies are stated. The last section concludes with some comments.

1.2 Concept of Green Technology

The term "green" denotes life, harmony, stability, and neutralization of nega-
tivity. The concept of green is in the heart of creation. Everything around
us is predominantly green. We depend on the green herbs, grass, trees, etc.
for life [2]. The term "technology" refers to the application of knowledge for
practical purposes. Green technology helps to reduce negative effects on
the environment while improving productivity, efficiency, and operational
performance of a given technology. The main goal of green technologies is
to meet the needs of society in the way that avoids depleting or damaging
natural resources on earth. The key components of green technologies are
recycling, environmental remediation, and renewable energy sources.

- *Recycling:* Green technology helps manage and recycle waste material. Recycle objects are made of glass, metal, paper, and plastic.
  These materials are reusable and should be recycled to prevent further depletion of the earth's resources.
- *Environmental remediation:* This involves removing contaminates
  from the soil, air, and water. It is the removal of pollutants or contaminants for the general protection of the environment.
- *Renewable energy sources:* Green technology includes the conversion
  of renewable resources to useful energy.

The four pillars of green technology are [3] (1) energy harvesting—this
seeks innovative ways to extract useful energy from waste by-products, as
well as to develop new technologies to maximize the harnessing of energy;
(2) environment—all human activities have an impact on the environment
and we must - minimize the impact; (3) economy—enhancing the national economic development through the use of technology will assist us in building a
strong and vibrant local community; and (4) social—we improve the quality of
life for all and emphasize the importance of individual well-being, including
full access to effective healthcare, housing, food, and education. Thus, green
technology addresses social, economic, and environmental values.

1.3 Going Green

Going green can help us come out of the present tough environmental problem.
Easy ways to be greener include [4]:

- Implement recycling in the office
- Remember: Reduce, reuse, recycle, repair, and think

- Make sure employees are aware, if recycling already exists
- Educate employees on what can and cannot be recycled (cardboard, plastics, glass fluorescents, and IT equipment)
- Purchase recycled paper
- Install water hippos in toilets (a device that sits in the cistern of the toilet and reduces water use with each flush) or use low-flow toilets
- Encourage employees for carpooling or using bike to work
- Turn off technology and/or appliances when not in use
- Buy LED, CFL (compact fluorescent bulbs) or other long-lasting bulbs
- Try to repurpose old technology
- Have heating and air conditioning on when needed, make sure it's not going non-stop
- Go paperless
- Do more things electronically (e.g., voting, filing taxes, and various tickets, and meetings)
- Ask boss about possibility of distant working
- Get rid of screen savers and allow products to go into sleep mode
- Order things online or walk or use e-bike, instead of using up fuel
- Try to repair products instead of replacing it entirely
- Buy a hybrid/electric car
- Use power-saving modes for maximum efficiency on all devices

These simple ways can make a significant impact.

1.4 Examples of Green Technology

Green technologies and practices are those which make an establishment's production processes more environment-friendly. Several companies world-wide have committed to establishing green business and manufacturing practices. These include IBM, Dell, Cisco, Hewlett-Packard, Johnson & Johnson, Intel, Nike, Wells Fargo, and Staples.

Green technologies and practices have been applied in several areas including green energy, green chemistry, green nanotechnology, and green buildings.

- *Green energy:* This is perhaps the most urgent use of green technology. Energy is being conserved through the use of green technology. Currently, nonrenewable resources make up 80% of the world's energy requirements, but they are not sustainable. Renewable energy sources include water, biomass, wind, solar, and geothermal. For example, a solar cell converts the energy in light into electrical energy. The development of green energy is the highest priority in many scientific endeavors [5].

- *Green chemistry:* This is also known as sustainable chemistry. It is a philosophy of chemical research and engineering that encourages the design of products and processes that minimize the use and generation of hazardous substances. Green chemistry applies to organic chemistry, inorganic chemistry, biochemistry, analytical chemistry, and physical chemistry [6].

- *Green nanotechnology:* This is one of the latest in green technologies. Nanotechnology involves the manipulation of materials at the atomic or nanoscale. Green nanotechnology is the application of green chemistry and green engineering principles to nanotechnology. Materials are manipulated in ways that will transform the manufacturing industry.

- *Green buildings:* The main benefit of building green is reducing a building's impact on the environment and significantly improving building performance. Using green technologies in building construction not only benefits the environment, also they can produce economically attractive buildings that are healthier for the occupants as well [7]. Using green roofs improved the energy performance of buildings because they provide higher thermal inertia, shading, and absorption of solar energy. Green buildings have the potential to substantially reduce energy consumption.

Other applications of green technology include water and waste management, green IT, green manufacturing, green business, green economics, green marketing, green supply chain, green logistics, transportation, smart grid, agriculture, construction industry, and water supply.

1.5 Advantages and Disadvantages

The advantages of green technologies are often stated in comparison to conventional and more polluting technologies. The advantages of green technology include [8, 9]:

1. Does not emit anything detrimental into atmosphere
2. Brings economic profits to certain areas
3. Needs less maintenance

4. Uses renewable natural resources that never depletes
5. Slows the impacts of global warming by reducing CO_2 emissions
6. Ensures maximum utilization of IT resources in the enterprise
7. Diminishes the number of malignant wastes to the atmosphere
8. Protects our planet from global warming

The disadvantages of green technology include [8]:

1. High implementing costs
2. Lack of information
3. No known alternative chemical or raw material inputs
4. No known alternative process technology
5. Uncertainty about performance impacts
6. Lack of human resources and skills

1.6 Benefits and Challenges

The most important benefit in applying green technology is enhancing the quality of life by ensuring a more sustainable environment. Other benefits of green technology includes recycling waste material, purifying of water, purifying the air, conserving energy, and rejuvenating ecosystems. Adoption of green technology can enhance a company's environmental reputation. Green technology is one of the fastest growing employment sectors and can generate job opportunities for those who are passionate about conserving the environment.

Green technologies face some challenges. They are generally more expensive than the established baseline technologies they aim to replace, because they have to account for the environmental costs. Selling green technologies has not been an easy task because it requires new paradigms of appreciation: cause consequence, cost benefit, etc. No change comes without some pain. A monopoly may continue to use a dirty production technology over a known socially superior green technology.

1.7 Conclusion

Green technology deals with using science and technology to protect the environment as well as curb the negative impacts of human involvement. It has emerged as an important trend and development in the 21st century.

Its importance has increased worldwide since the turn of the century. Its development will lead to global and sustainable powers that will impact our economics, societies, cultures, and way of life. Several business establishments have used at least one green technology or practice in order to make their production processes more environment-friendly. More information on green technologies can be found in the book in [10–13].

References

1. M. N. O. Sadiku et al., "Green technology," *International Journal of Trend in Scientific Research and Development*, vol. 3, no. 1, November–December, 2018, pp. 1137–1139.
2. J. O. Odigure, "Green technology: A veritable tool for achieving technological transformation and diversification of the nation's economy," *7th Biennial National Engineering Conference*, Bida, Nigeria, September 2016.
3. M. Bhardwa and Neelam, "The advantages and disadvantages of green technology," *Journal of Basic and Applied Engineering Research*, vol. 2, no. 22, October–December, 2015, pp. 1957–1960.
4. S. Mueller, "Green technology and its effect on the modern world," *Bachelor's Thesis*, Oulu University of Applied Sciences, Spring 2017.
5. M. N. O. Sadiku, S. R. Nelatury, and S. M. Musa, "Green energy: A primer," *Journal of Scientific and Engineering Research*, vol. 5, no.7, 2018, pp. 336–339.
6. M. N. O. Sadiku, S. M. Musa, and O. M. Musa, "Green chemistry: A primer," *Invention Journal of Research Technology in Engineering and Management*, vol. 2, no. 9, September 2018, pp. 60–63.
7. G. D. Soni, "Advantages of green technology," *International Journal of Research— GRANTHAALAYAH*, vol. 3, September 2015.
8. A. Iravani, M. H. Akbari, and M. Zohoori, "Advantages and disadvantages of green technology; goals, challenges and strengths," *International Journal of Science and Engineering Applications*, vol. 6, no. 9, 2017, pp. 272–284.
9. A. Mehdialiyev and O. Mazanova, On some problems of the creation and development of green technologies in Azerbaijan," https://www.researchgate.net/publication/271463949_On_some_problems_of_the_creation_and_development_of_green_technologies_in_Azerbaijan
10. S. B. Billatos and N. A. Basaly, *Green Technology and Design for the Environment*. Taylor & Francis, Washington, DC, 1997.
11. R. Singh and S. Kumar (eds.), *Green Technologies and Environmental Sustainability*. Springer, Singapore, 2017.
12. C. Y. Foo, T. A. Aziz, and M. Kamal (eds.), *Green Technologies for the Oil Palm Industry*. Springer, Singapore, 2019.
13. H. H. Ngo et al. (eds.), *Green Technologies for Sustainable Water Management*. American Society of Civil Engineers, Virginia, 2016.

2

Green Energy

The only real mistake is the one from which we learn nothing.

—John Powell

2.1 Introduction

Energy may be defined as the capacity to do work. There are various types of energy: electrical energy, chemical energy, nuclear energy, thermal energy, gravitational energy, potential energy, etc. Energy is crucially important in the economic and social development of any society. The use of energy is evident in our everyday lives. We need energy for lighting, heating, cooling, transport, communication systems, domestic appliances, and battery-powered devices to mention but a few. The conventional type of energy is fossil-based energy, which generally includes coal, petroleum, natural gas, etc. Energy that comes from these conventional means is called "brown energy." Another type of energy is "green energy," which is a clean source of energy with a lower environmental impact compared to conventional sources [1].

Green energy, which is sometimes called renewable or sustainable energy, comes from natural sources such as wind, water, and sunlight. It is often called "clean" because it produces no pollutants. It is an attractive option because it provides a clean and earth-friendly alternative to traditional energy. Different types of green energy sources are illustrated in Figure 2.1 [2].

Concerns about climate change and global warming are driving increasing renewable energy legislation. Green or renewable energy is a form of energy that does not contribute to climate change or global warming. It uses energy sources that are continually replenished by nature. It can fulfill the need for energy while saving the planet. Renewable energy is derived from natural resources such as sunlight, wind, tides, and geothermal heat that are replenished constantly. The use of green energy would reduce environmental pollution and lead to energy independence and an electrical grid that is much more reliable, secure, efficient, and greener. As Sherrod Brown (the US senator) rightly said, "Green energy is an environmental strategy, a national security strategy, an economic strategy. Investing in its development and

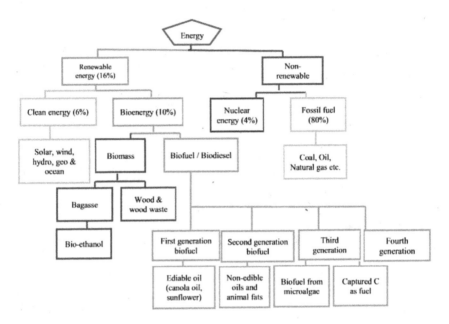

FIGURE 2.1
Green sources [2].

production is both right and smart. Failing to invest in it is a risk to the future of our nation and our planet" [3].

This chapter briefly reviews major types of renewable energy and their usage. It begins by presenting the conventional or brown energy. It covers different types of green energy sources: solar, wind, geothermal, and biomass. It presents some usages or applications of green energy. It also discusses the issue of renewable energy efficiency and addresses the benefits and challenges of green energy. The last section concludes the chapter.

2.2 Conventional Energy

Energy resources help in creating wealth and improving living standards. The availability and affordability of energy sources are crucial for the overall economic development of a nation. Conventional sources of energy are nonrenewable fossil fuels (coal, oil, gas, petroleum, etc.) and nuclear power [4]. Fossil fuels have been used as a common source of energy for centuries. Petroleum springs and coal mines are not inexhaustible and are rapidly diminishing in many places. A lot of attention has been focused on the environmental impacts of conventional energy sources, particularly fossil fuels.

Energy that comes from these conventional means is called "brown" (carbon-intensive) energy. Grid power is a mixture of brown and renewable energies. It is regarded as a kind of brown energy source because utilities produce much of their power by burning carbon-intensive fossil fuels, such as coal and natural gas. Most brown energy sources (such as fossil fuel power, gas turbine, coal, and oil) are able to adjust their output power on demand through tuning the power generator. If brown energy must be used, the scheduler selects times when it is cheap.

To reduce the harmful effects of brown energy, green energy sources need to be applied. Although each country has both traditional power plants and green energy plants, there is a need for a change in the energy consumption pattern. Earth resources are limited and are increasingly depleting. This compels governments and environmental activists all over the world to emphasize the need to switch from conventional resources to alternative green sources [5]. Green energy is the type that produces few externalities for environment. It uses electricity and gas made from renewable sources. As the costs of brown or nonrenewable energy grow, renewable energy becomes more widely used.

2.3 Types of Green Energy

Green energy is generated from sources such as solar, wind, geothermal, and biomass. There are different forms of green energy depending on the sources [6, 7]. The sun plays a crucial role in most types of renewable energy, since they depend on it one way or the other.

- *Solar energy:* This involves capturing the sun's energy with photovoltaic (PV) cells. The solar cells absorb the solar radiation from the sun and convert the energy into direct current electricity. This energy can be collected and converted in different ways. It can be harnessed through a range of technologies like solar heating and photovoltaics. Solar energy generation depends on various factors such as temperature, solar intensity, and geolocation of the solar panels. Solar energy is the fastest growing renewable technology, while solar photovoltaics is the largest renewable energy employer. For example, a building can be constructed to incorporate a solar hot water, cooling, or ventilation system. Solar energy can supply solar heat to houses and industrial processes.

- *Wind energy:* Wind energy is produced by wind turbines with rotating blades capturing the wind flow and harnessing the wind's kinetic energy to generate electricity. It requires extensive area coverage to produce significant amounts of energy. For example, wind turbines may be used to generate electricity as a supplement to a company's existing power supply. Wind is most likely the safest form of green energy in terms of its overall ecological impacts. Wind energy is the most advanced renewable energy.

- *Hydroelectric energy:* Moving water is responsible for powering turbines and generating hydroelectric power. The turbines are connected to generators that harness the mechanical energy from the water currents and convert it into hydroelectricity. This is often regarded as the largest source of renewable energy because it provides more than 97% of all the electricity generated by renewable sources worldwide. Strictly speaking, hydropower is not renewable because it has the largest environmental impacts partly due to the need to construct dams that block animal migration and disrupt river flows. For example, small towns can harness the energy of local rivers by building hydroelectric power systems.

- *Geothermal energy:* As the name implies, geothermal energy is heat energy from the earth itself. The temperature of the earth steadily increases with depth. Geothermal power plants harness the heat sources to produce electricity, which is cost-effective, reliable, sustainable, and nonpolluting or eco-friendly. A major challenge with this energy source is that plants are expensive to build. For example, geothermal energy may be used for heating/cooling office buildings or manufacturing plants. Unlike solar and wind energies, which are intermittent, geothermal energy can be generated 24/7.

- *Biomass:* This is produced when organic wastes decay. This waste can be converted to fuel through combustion for the generation of electricity. Biomass is mankind's original source of energy. The most popular form is burning trees for cooking and warmth. Geothermal and biomass power plants may require water for cooling. For example, farm operations can convert waste from livestock into electricity. Unlike solar and wind, biomass power is dispatchable, that is, it can be turned on and off.

Other types of green energy include energy from tides, hydrogen, and fuel cells. All these are produced with the same goal in mind, which is to save the planet. Countries such as Iceland and Norway have reached 100% renewable energy generation, that is, all their electricity is generated using renewable energy. Renewable energy technologies are getting cheaper due to mass production and market competition.

2.4 Applications

Rapid advances in green energy technologies are improving the efficiency of generating electricity using renewable sources, and also driving down the cost of deploying a green power system. Here we present some usages or applications of green energy.

- *Smart home:* Home energy consumption, such as electricity, heating, and cooling, has been an important environmental and economic issue for decades. Constraints such as renting, safety, and unsupportive household members affect energy use and energy saving behaviors of customers [8]. A smart home provides optimum living conditions required naturally. A green renewable energy source (such as a solar panel) has been utilized in generating power for all the smart appliances used to sustain the smart home. Solar heat energy has been used to generate hot water and do the cooking. Using the green energy source in smart homes can reduce energy cost and minimize wastage of energy [9]. A typical solar energy generation at home is shown in Figure 2.2 [10].

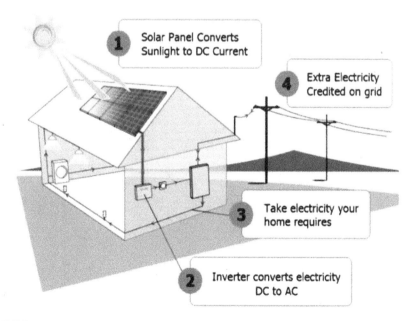

FIGURE 2.2
A typical example of solar energy generation in a home [10].

- *Businesses:* Reducing energy usage is not limited to household. For businesses, it is important to effectively reduce electricity consumption and environmental pollution. Switching to green energy can actually bring many different benefits to a business. The benefits include lower energy bills, boosting public relations, creating jobs, and great return on investment [11].

- *Data centers:* Data centers are known for consuming an enormous amount of electricity. Mega data centers (such as those of Apple, Microsoft, and Google) have emerged due to the soaring demand for IT services. Data center operators are constantly under pressure to minimize the carbon footprint. To achieve this, powering data centers by on-site generation of renewable energy is required. Renewable energy integration lowers the cost of designing fault-tolerant distributed data centers with reduced carbon footprint [12]. To reduce costs and environmental impacts, modern data center operators, such as those of Google and Apple, are beginning to integrate renewable energy sources into their power supply. Sometimes, data centers can "bank" green energy in batteries or on the grid. Different options for green and brown energy sources for green data centers are illustrated in Figure 2.3 [13].

- *Mobile networks:* These are among the major energy guzzlers. The growing energy consumption leads to a significant rise of carbon footprints. Therefore, greening mobile networks is becoming a necessity for economic and environmental sustainability. Green energy is a promising energy alternative for future mobile networks [14].

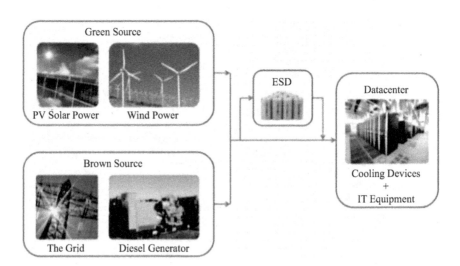

FIGURE 2.3
Different options for green and brown energy sources for green data center [13].

- *Cellular networks:* An increase in the number of mobile users and their diverse data applications is compelling cellular network operators to install more base stations (BSs). Concerns about the increasing number of BSs with high energy consumption have prompted cellular operators to deploy renewable energy sources in BSs. This helps reduce the on-grid consumption and operational expenditure. Powering cellular networks with renewable energy sources is a promising alternative for reducing global carbon footprint [15].

Other applications include cognitive radio networks, cyber-physical systems, and battlefield.

2.5 Renewable Energy Efficiency

Improving energy efficiency at homes, businesses, offices, schools, governments, and industries is a cost-effective way to address the challenges of high energy prices and energy. Sometimes energy efficiency may mean changing behaviors, such as drying clothes on a clothesline instead of a dryer. Better ways to produce solar photovoltaic panels include lowering panel production cost.

Governments can promote energy efficiency by setting energy efficiency standards. Products that meet high-efficiency standards should receive the Energy Star label. Efficiency labels inform consumers about the energy efficiency of different products.

Unfortunately, even with informative labels, some consumers do not purchase high-efficiency products because the up-front costs may be higher than regular [16].

2.6 Benefits

Green energy offers a number of benefits to businesses and institutions. Renewable energy has the potential to lift the poorest nations to new levels of prosperity since it is rapidly becoming more efficient and cheaper. It is particularly suitable for developing nations. Producing renewable energy locally can benefit rural and remote areas [17].

In a sense, green energy is unlimited since supplies are continually replenished through natural processes. Choosing green power is a prudent step toward more sustainable operations and practices and a demonstration of environmental stewardship. Green energies tend to have much lower emissions than other sources, such as natural gas or coal. The renewable energy industry is more labor-intensive and supports thousands of jobs. It is providing stable and affordable electricity [18]. Many homeowners can sell excess solar or wind energy to their utility companies. This way, they can pay off their energy investments quickly.

2.7 Challenges

The most significant challenges to the widespread implementation of renewable energy are seen to be mainly social and political, not technological or economic. The key barriers to adoption of green energy include climate change denial, the fossil fuels lobby, political inaction, higher generation cost, higher market price, unsustainable energy consumption, and outdated energy infrastructure. Renewable energy technology has sometimes been regarded as expensive by critics, and affordable only in the affluent developed world. But renewable energy can be suitable for developing countries as well. It can contribute to poverty reduction by providing the energy needed for creating businesses and jobs [16].

Renewable energy from sources such as wind power and solar power is sometimes criticized for being variable and not available 24/7. Hydroelectric power generators can disrupt river ecosystems both upstream and downstream from the dam.

Economic constrain is also a challenge, as substantial investment is needed to implement renewable energy. Investment in green energy depends on the availability of finance. Developed economies dominated the financing of renewable energy. In developing economies, investing in green energy technology is difficult because of the high cost of financing and also because obtaining financing at affordable rates is a major challenge. Political will is necessary to accomplish the transition from brown to green energy.

2.8 Conclusion

Globalization has led the modern society toward green sources of energy, which are clean sources of energy with a lower environmental impact compared to traditional conventional energy sources. Green is becoming fashionable today. Green energy has attracted much attention across the globe due to the fact that it is nonpolluting and more environment-friendly. Renewable energy systems are rapidly becoming more efficient and cheaper. The market for renewable energy will keep growing. The renewables are the energy sources of the future.

There is an incentive to use 100% renewable energy to help confront issues related to climate change, energy security, and the escalation of energy costs. Each country should have green energy as an important component of energy planning. Government policies are important in ensuring that the energy sector produces sustainable energy. Policy instruments such as taxes, regulations, and subsidies can stimulate the adoption of green energy technologies [19].

Millions of renewable energy jobs will be available for qualified workers over the coming decade. Training and education are essential in preparing workers to take advantage of these opportunities. More education about green energy is necessary for the general public, students, and engineers to be aware of the new field. For more information about green renewable energy, see references of books [16, 20–27] and journals devoted exclusively to green energy:

- *International Journal of Green Energy*
- *Energy*
- *Green Energy & Environment*
- *Journal of Fundamentals of Renewable Energy and Applications.*

References

1. M. N. O. Sadiku, S. R. Nelatury, and S. M. Musa, "Green energy: A primer," *Journal of Scientific and Engineering Research*, vol. 5, no. 7, 2018, pp. 336–339.
2. S. A. R. Khan and D. Qianli, "Does national scale economic and environmental indicators spur logistics performance? Evidence from UK," *Environmental Science and Pollution Research*, vol. 24, no. 34, December 2017, pp. 26692–26705.
3. S. Brown, "A case for green energy manufacturing," *New Solutions*, vol. 19, no. 2, 2009, pp. 135–137.
4. D. Lidgate, "Green energy?" *Engineering Science and Education Journal*, vol. 1, no. 5, October 1992, pp. 221–227.
5. C. Bhowmik, S. Bhowmik, and A. Ray, "Social acceptance of green energy determinants using principal component analysis," *Energy*, vol. 160, 2018, pp. 1030–1046.
6. "Types of green," https://www.igsenergy.com/your-energy-choices/green-energy/types-of-green/
7. "7 types of renewable energy to support commercial sustainability," http://businessfeed.sunpower.com/lists/7-types-of-renewable-commercial-energy
8. T. Dillahunt et al., "It's not all about 'green': Energy use in low-income communities," *Proceedings of the 11th international conference on Ubiquitous computing*, Orlando, Florida, September–October 2009, pp. 255–264.
9. D. Nag et al., "Green energy powered smart healthy home," *Proceedings of the 8th Annual Industrial Automation and Electromechanical Engineering Conference*, August 2017, pp. 47–51.

10. "Evolution solar," http://kingaroy.evolutionsolar.com.au/solar-power/solar-power-explained-evolution-solar-kingaroy/
11. "9 ways businesses can benefit from renewable energy," https://www.conserve-energy-future.com/9-ways-businesses-can-benefit-renewable-energy.php
12. R. Tripathi, S. Vignesh, and V. Tamarapalli, "Optimizing green energy, cost, and availability in distributed data centers," *IEEE Communications Letters*, vol. 21, no. 3, March 2017, pp. 500–503.
13. F. Kong and X. Li, "A survey on green-energy-aware power management for datacenters," *ACM Computing Surveys*, vol. 47, no. 2, November 2014.
14. T. Han and N. Ansari, "Powering mobile networks with green energy," *IEEE Wireless Communications*, February 2014, pp. 90–96.
15. A. Jahid, A. B. Shams, and F. Hossain, "Green energy driven cellular networks with JT CoMP technique," *Physical Communication*, vol. 28, 2018, pp. 58–68.
16. D. Timmons, J. M. Harris, and B. Roach, *The Economics of Renewable Energy*. Medford, MA: Global Development and Environment Institute, 2014.
17. "Renewable energy," Wikipedia, the free encyclopedia, https://en.wikipedia.org/wiki/Renewable_energy
18. "Benefits of renewable energy use," https://www.ucsusa.org/clean-energy/renewable-energy/public-benefits-of-renewable-power#.WxBWYkOo7nMContents
19. M. Woerter et al., "The adoption of green energy technologies: The role of policies in Austria, Germany, and Switzerland," *International Journal of Green Energy*, vol. 14, no. 14, 2017, pp. 1192–1208.
20. X. Li (ed.), *Green Energy: Basic Concepts and Fundamentals*. London: Springer, 2011.
21. D. Elliott (ed.), *Sustainable Energy: Opportunities and Limitations*. London: Palgrave Macmillan, 2007.
22. E. Jeffs, *Greener Energy Systems: Energy Production Technologies with Minimum Environmental Impact*. Boca Raton, FL: CRC Press, 2012.
23. E. Jeffs, *Green Energy: Sustainable Electricity Supply with Low Environmental Impact*. Boca Raton, FL: CRC Press, 2017.
24. U. Aswathanarayana, T. Harikrishnan, and T. S. Kadher-Mohien, *Green Energy: Technology, Economics and Policy*. Boca Raton, FL: CRC Press, 2010.
25. J. H. Appelman, A. Osseyran, and M. Warnier, *Green ICT & Energy: From Smart to Wise Strategies*. Boca Raton, FL: CRC Press, 2013.
26. J. Byrne and Y. D. Wang, *Green Energy Economies: The Search for Clean and Renewable Energy*. Boca Raton, FL: CRC Press, 2014.
27. Y. Demirel, *Energy: Production, Conversion, Storage, Conservation, and Coupling*, 2nd ed. London: Springer, 2016.

3

Green Chemistry and Engineering

The seven C's of success: Courage, commitment, confidence, creativity, common sense, character, communion with God.

—**Matthew N. O. Sadiku**

3.1 Introduction

The chemical industry is perhaps the most successful and diverse sector of the manufacturing industry. The globalization and digitalization of value chains change production and business models in the chemical industry. Chemical industry manufactures chemical products that are used in healthcare, agriculture and food, electronics, clothing, and transportation. These products undoubtedly have contributed enormously to the quality of our lives and increased longevity of the human race. However, there is unprecedented social, economic, and environmental pressure on the chemical industry to make chemical processes and products more sustainable and environmentally compatible [1].

Concerns about global climate change, shortages of water, depletion of materials, degradation of the environment, and soaring energy prices are driving new priorities and expectations. There has also been serious demand for green and renewable chemicals around the globe. To meet these demands, engineers and scientists worldwide are leading the charge and crafting strategies to address them. Engineering is a profession that solves problems confronting our civilization. Green engineering (GE) is the discipline that promotes the idea of making things better for the environment right from the beginning. Every manufacturer is now under scrutiny for how environmental friendly its processes and products are.

GE is the design of processes and products that minimize pollution, promote sustainability, and protect human health without sacrificing economic viability and efficiency. It is the process of using hardware and software technologies to reduce our impact on the environment. It also involves designing materials, processes, and devices over the entire life cycle of a product [2]. GE advocates the idea that decisions to protect human health and the environment can have the greatest impact and cost-effectiveness when applied early.

Green chemistry (GC) and GE use scientific knowledge to reconcile the need for chemical production and the desire to reduce or eliminate the use of hazardous substances. They have the potential to benefit both the economic and environmental agendas [3]. It is evident to everyone that life without chemistry and chemical is inconceivable, while, on the other hand, the potential risks and impacts to the environment associated with chemical production and chemical products cannot be ignored. These societal and environmental obligations require that all chemical companies should be accountable for their actions [4].

GE financially and technologically designs products and processes in a manner that simultaneously decreases the amount of pollution and minimizes exposures to potential hazards. GE is not actually an engineering discipline, but an overarching engineering framework for all design disciplines. It is closely related to environmental engineering [5]. Thus, this branch of engineering has variably been called "environmental," "chemical," "ecological," "clean," "sustainable," and "green."

This chapter provides a brief introduction to GC and GE. It begins by illustrating how the 12 principles of GC and GE can be used to aid in design. A range of illustrative common examples or applications of GC and GE are presented. Some benefits and challenges of these technologies are addressed. The last section concludes the chapter.

3.2 Green Chemistry Principles

With mounting concerns over the state of our planet, there is continuing demand that chemists and chemical engineers should develop greener chemical processes and products. In the 1990s, with the growing awareness of the hazardous impacts of the chemical industry, the GC revolution was launched by American chemists Paul T. Anastas and John Warner. GC is the kind of chemistry that seeks to minimize pollution, conserve energy, and promote environmentally friendly production.

The concept of GC is relatively a new idea. GC involves the development of chemical products and processes that reduce or eliminate the use of hazardous substances. It is cleaner and smarter chemistry. GC is not confined to industrial sector. It applies to all areas of chemistry including organic chemistry, inorganic chemistry, biochemistry, analytical chemistry, and physical chemistry. It is an interdisciplinary area drawing knowledge from chemists, chemical engineers, toxicologists, and ecologists. It has evolved from being academic research effort to become a practice supported by academia, industry, and government.

GC is an integral part of GE since it provides the foundation on which to build GE. Sustainable engineering transforms current engineering practices to those that promote sustainability. The relationship between GC, GE, and sustainability is shown in Figure 3.1 [6]. Combining GC with GE

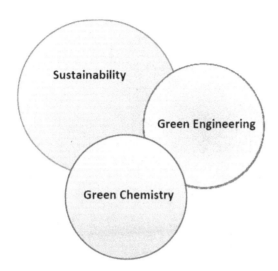

FIGURE 3.1
Relationship between green chemistry, green engineering, and sustainability [6].

at the earliest design stages will maximize efficiency, minimize waste, and increase profitability [6]. From design to disposal, GE is finding ways to balance environmental compatibility with economic profitability.

In 1998, Paul Anastas and John C. Warner published 12 principles to guide the practice of GC. The principles address a wide range of ways to reduce the environmental and health impacts of chemical production. They provide an early conception of what would make a greener chemical or product. The 12 principles are illustrated in Figure 3.2 and explained as follows [7–9].

1. *Prevention:* Preventing waste is better than treating or cleaning up waste after it is created.
2. *Atom economy:* Synthetic methods should try to maximize the incorporation of all materials used in the process into the final product.
3. *Less hazardous chemical syntheses:* Synthetic methods should avoid using or generating substances toxic to humans and/or the environment.
4. *Designing safer chemicals:* Chemical products should be designed to achieve their desired function while being as nontoxic as possible.
5. *Safer solvents and auxiliaries:* Auxiliary substances should be avoided wherever possible and as nonhazardous as possible when they must be used.
6. *Design for energy efficiency:* Energy requirements should be minimized, and processes should be conducted at ambient temperature and pressure whenever possible.
7. *Use of renewable feedstocks:* Whenever it is practical to do so, renewable feedstocks or raw materials are preferable to nonrenewable ones.

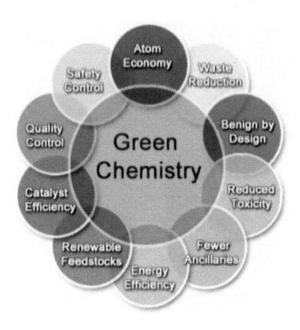

FIGURE 3.2
The principles of green chemistry [9].

8. *Reduce derivatives:* Unnecessary generation of derivatives—such as the use of protecting groups—should be minimized or avoided if possible; such steps require additional reagents and may generate additional waste.

9. *Catalysis:* Catalytic reagents that can be used in small quantities to repeat a reaction are superior to stoichiometric reagents (ones that are consumed in a reaction).

10. *Design for degradation:* Chemical products should be designed so that they do not pollute the environment; when their function is complete, they should break down into non-harmful products.

11. *Real-time analysis for pollution prevention:* Analytical methodologies need to be further developed to permit real-time, in-process monitoring and control *before* hazardous substances form.

12. *Inherently safer chemistry for accident prevention:* Whenever possible, the substances in a process, and the forms of those substances, should be chosen to minimize risks such as explosions, fires, and accidental releases.

Winterton proposed 12 more green principles [10]:

1. Identify and quantify by-products.
2. Report conversions, selectivities, and productivities.

3. Establish full mass balance for the process.

4. Measure catalyst and solvent losses in air and aqueous effluent.

5. Investigate basic thermochemistry.

6. Anticipate heat and mass transfer limitations.

7. Consult a chemical or process engineer.

8. Consider the effect of overall process on choice of chemistry.

9. Help develop and apply sustainability measures.

10. Quantify and minimize the use of utilities.

11. Recognize where safety and waste minimization are incompatible.

12. Monitor, report, and minimize the laboratory waste emitted.

These principles are being put into practice in all walks of science since they provide a recipe for an efficient and safe environment. The principles represent applying the ethics of thrift to the practice of chemistry, thereby serving economic, social, and environmental goals. With these principles, GC aims to reduce or eliminate waste in the manufacture of chemicals and its allied products.

3.3 Green Engineering Principles

Anastas and Zimmerman [11] organized an overview of developments in GE into 12 principles. Their 12 Principles of GE have been proposed as a framework for examining existing products as well as to evaluate new product designs. GE may be regarded as the incorporation of these 12 principles into engineering practices. The principles are stated as follows [11, 12]:

1. *Inherent rather than circumstantial:* Designers need to strive to ensure that all materials and energy inputs and outputs are as inherently nonhazardous as possible.

2. *Prevention instead of treatment:* It is better to prevent waste than to treat or clean up waste after it is formed.

3. *Design for separation:* Separation and purification operations should be designed to minimize energy consumption and materials use.

4. *Maximize efficiency:* Products, processes, and systems should be designed to maximize mass, energy, space, and time efficiency.

5. *Output pulled versus input pushed:* Products, processes, and systems should be "output pulled" rather than "input pushed" through the use of energy and materials.

Principles of Green Engineering		Principles of Green Chemistry	
I -	Inherently non-hazardous and safe	P -	Prevent wastes
M -	Minimize material diversity	R -	Renewable materials
P -	Prevention instead of treatment	O -	Omit derivatization steps
R -	Renewable material and energy inputs	D -	Degradable chemical products
O -	Output-led design	U -	Use safe synthetic methods
V -	Very simple	C -	Catalytic reagents
E -	Efficient use of mass, energy, space & time	T -	Temperature, Pressure ambient
M -	Meet the need	I -	In-Process Monitoring
E -	Easy to separate by design	V -	Very few auxiliary substances
N -	Networks for exchange of local mass and energy	E -	E-factor, maximize feed in product
T -	Test the life cycle of the design	L -	Low toxicity of chemical products
S -	Sustainability throughout product life cycle	Y -	Yes it's safe

FIGURE 3.3
Abbreviated principles of green engineering and green chemistry [13].

6. *Conserve complexity:* Embedded entropy and complexity must be viewed as an investment when making design choices on recycle, reuse, or beneficial disposition.

7. *Durability rather than immortality:* Targeted durability, not immortality, should be a design goal.

8. *Meet need, minimize excess:* Design for unnecessary capacity or capability (e.g., "one size fits all") solutions should be considered a design flaw.

9. *Minimize material diversity:* Material diversity in multicomponent products should be minimized to promote disassembly and value retention.

10. *Integrate material and energy flows:* Design of products, processes, and systems must include integration and interconnectivity with available energy and materials flows.

11. *Design for commercial "Afterlife":* Products, processes, and systems should be designed for performance in a commercial "afterlife."

12. *Renewable rather than depleting:* Material and energy inputs should be renewable rather than depleting.

Green engineers who design products allow these basic principles to govern how they perform their work. The principles have been put to use in fostering sustainability. The abbreviated 24 Principles of GC and GE, with the mnemonic "IMPROVEMENTS PRODUCTIVELY," are illustrated in Figure 3.3 [13].

3.4 Drivers for Green Adoption

Activities in the field of GC and GE are increasing rapidly. While GE is concerned with the design, discovery, commercialization, and use of processes and products, GC is a subset of GE, which covers the design of chemical processes and products. GC and GE are emerging areas that come under a larger area of sustainable development [14].

All scientists and engineers can employ GC and GE principles regardless of the specific area of specialization. GE encompasses common measurements such as power quality, consumption, and emissions from vehicles and factories. Factors that are motivating companies to go green include environmental legislation, rising waste-disposal costs, and corporate image.

GC may be regarded as a design philosophy since it requires incorporating at the design stage of a of a new chemical or proces. GC provides metrics to evaluate the efficiency, environmental, and health impacts of new technology [15].

3.5 Metrics

To measure the environmental impact of chemical processes and track GC progress, some tools are needed. GC can be defined using of metrics, which are used to quantify greener processes and products. In other words, the metrics are used to determine how green is green [16]. Such an effort would require the collaboration of chemists, policy makers, educators, and businesses [17]. These metrics include ones for mass, energy, hazardous substance reduction, and life cycle environmental impacts.

Source reduction and higher recycling are unlikely to take place without government intervention.

3.6 Applications

The main goal of GE is meeting the needs of society in ways without depleting natural resources on the planet. GE is a way to make many of the things people use in everyday life more efficient, safe, and long-lasting. Thus, GE is an environmentally friendly engineering. Engineers who want to lower the emissions of their products and develop devices that consume less energy need GE.

GE is all around you. Examples include solar cells, reusable water bottles, green buildings, and newer cars that run more cleanly. More detailed examples are presented as follows.

- *Pharmaceutical manufacturing:* The main goal of GC is to reduce or eliminate waste in the manufacture of drugs. The pharmaceutical industry was among the first to recognize the value of GC and GE. The pharmaceutical industry produces medicines that allow patients to be healthy, be productive, and live long. It is committed to produce medicines with minimal environmental impact. There are opportunities to reduce the environmental footprint and economic cost of pharmaceutical manufacturing. GE represents a practical transformation of the industry [17].

- *Green nanotechnology:* In the context of nanotechnology, green innovation is aimed at processes for the production of products that are safe, energy efficient, and minimize greenhouse emissions. Green nanotechnology is about manufacturing processes that are environmentally sustainable [18]. It has the goal of producing nanomaterials and products without harming the environment. For example, nanoscale catalysts can cause chemical reactions to be more efficient and less wasteful.

- *Green building:* Green building is becoming common as more owners feel a responsibility to build sustainably. Green building consists of water recycling, solar power installations, cooling systems, and energy-efficient window systems. It provides greener solutions to normal building processes that conserve water, save energy, and provide cleaner indoor air, while also reducing utility bills. It is designed to independently use renewable energy. Sustainable building will become a common practice in the future [19].

- *Green energy:* Green or renewable energy is a form of energy that does not contribute to climate change or global warming. It uses energy sources that are continually replenished by nature. It comes from natural sources like wind, water, and sunlight. Green energy has attracted much attention across the globe due to the fact that it is nonpolluting and more environmentally friendly [20].

- *Leisure industry:* Chemistry provides inexpensive materials for golf, fishing, etc. Tourism activities can have negative impacts when the environment is unable to cope with the level of visitors. Tourism can contribute to the depletion of natural resources such as water, food, and energy. GC can contribute to the sustainability of leisure industry [21].

Other applications include green employment, green process engineering, and green products.

3.7 Benefits and Challenges

GC and GE are tools for achieving sustainability. It is sustainable in the sense that the technologies serve the needs of today's generation without endangering the ability of future generation to meet their own needs.

GC is beneficial to our planet. Adopting GC gives us an opportunity to create a safer laboratory environment. It provides greater efficiency in the chemical industry. It has the potential to improve the quality of different technologies used in chemical industry.

Going green can benefit us in the following ways [22]. Using sustainable and renewable resources for reactants and catalysts preserves resources for future generations. When less waste is produced in a reaction, there is less potentially hazardous material being released into the environment. It also prevents class actions lawsuits, which are expensive and damaging to a company's reputation.

Developing and implementing GC and GE requires investment. The cost of implementing GE over traditional solutions can be a deterrent. The disadvantages of GC and GE include high implementation costs, lack of information, uncertainty about performance impacts, and lack of human resources and skills [23].

The GC revolution presents a number of challenges to chemists and chemical engineers. Major challenges confronted by GC innovations include [24]: (1) economic and financial, (2) regulatory, (3) technical, (4) organizational, (5) cultural, and (6) definition and metrics. A large part of the chemical industry is capital-intensive. Large companies are slow and reluctant to switch to new technologies. It is also a challenge for industries to synthesize non-harmful products.

GE is complex by nature. Although engineers have learned to design systems with a modest degree of complexity, the need for increased complexity is growing.

In China, for example, the major impediments to implement GE seem to be competing priorities between economic growth and environmental protection. It still needs to train and educate its scientists and engineers [3].

Integrating GE into a nongreen curriculum can be challenging. Since GE is unique, teaching it must be carefully implemented in order to effectively present the material to students [25]. GC demands new standards for chemicals and chemical processes. In spite of these challenges, GC and GE are rapidly growing.

3.8 Conclusion

GC and GE are basically ways to make many of the things people use in everyday life more efficient, long-lasting, and eco-friendly. They involve creating healthy living environments that use natural resources wisely and do more with less. They form the framework for an eco-friendly design and commercialization. They assume that economic goals and environmental goals are compatible and can be achieved simultaneously. Companies are now realizing that it makes more sense to start designing for sustainability right from the start and that GE has positive effects on corporate profitability as well. GC has emerged as an important aspect of all chemistry and it is the future of chemistry. GC and GE are changing their practices to be more environmentally responsible.

Education is a crucial tool for realizing the new concept of GE, which is interdisciplinary in nature. GE principles are gaining attention in engineering education. It is being introduced to students at both the undergraduate and graduate levels in the United States and around the world. A GE course must be application oriented. The inclusion of GE tools and principles in engineering education will grow over the years [26–28]. More information about GC and GE can be found in books in References [10, 14, 29–36] and the journals exclusively devoted to them:

- *Green Chemistry*
- *Current Opinion in Green*
- *Sustainable Chemistry*
- *Journal of Green Engineering*
- *The Green and Sustainable Chemistry*
- *International Journal of Wastewater Treatment and Green Chemistry*
- *International Journal of Green Chemistry*
- *Green Chemistry for Sustainability*

References

1. J. H. Clark and D. Macquarrie (eds.), *Handbook of Green Chemistry and Technology*. Hoboken, NJ: Blackwell Science, 2002, p. vii.
2. M. N. O. Sadiku, S. R. Nelatury, and S. M. Musa, "Green engineering: A primer," *Journal of Scientific and Engineering Research*, vol. 5, no.7, 2018, pp. 20–23.
3. K. J. M. Matus, X. Xiao, and J. B. Zimmerman, "Green chemistry and green engineering in China: Drivers, policies and barriers to innovation," *Journal of Cleaner Production*, vol. 32, 2012, pp.193–203.
4. D. Shonnard, A. Kicherer, and P. Salin, "Industrial applications using BASF eco-efficiency analysis: Perspectives on green engineering principles," *Environmental Science & Technology*, vol. 37, no. 23, 2003, pp. 5340–5348.
5. "Green engineering," *Wikipedia*, the free encyclopedia, https://en.wikipedia.org/wiki/Green_engineering
6. "Is sustainable energy development possible?" Unknown website.
7. M. Kirchhoff, "Promoting green engineering through green chemistry," *Environmental Science & Technology*, vol. 37, October 2003, pp. 5349–5353.
8. P. T. Anastas and J. B. Zimmerman, "12 principles of green engineering," *Environmental Science & Technology*, vol. 37, no. 3, 2003, pp. 94A–101A.

9. M. N. O. Sadiku, S. M. Musa, and O. M. Musa, "Green chemistry: A primer," *Invention Journal of Research Technology in Engineering and Management*, vol. 2, no. 9, September 2018, pp. 60–63.

10. A. S. Matlack, *Introduction to Green Chemistry*, 2nd ed. Boca Raton, FL: CRC Press, 2010, p. vii.

11. M. Tobiszewski et al., "Green chemistry metrics with special reference to green analytical chemistry," *Molecules*, vol. 20, 2015, pp. 10928–10946.

12. J. A. Tickner and M. Becker, "Mainstreaming green chemistry: The need for metrics," *Current Opinion in Green and Sustainable Chemistry*, vol. 1, 2016, pp. 1–4.

13. "The importance of green chemistry," https://www.industrysearch.com.au/the-importance-of-green-chemistry/f/18149

14. R. R. Dupont, K. Ganesan, and L. Theodore, *Pollution Prevention: Sustainability, Industrial Ecology, and Green Engineering*, 2nd ed., chapter 3. Boca Raton, FL: CRC Press, 2017.

15. "12 principles of green engineering," https://www.acs.org/content/acs/en/greenchemistry/what-is-green-chemistry/principles/12-principles-of-green-engineering.html

16. N. Asfaw et al., "The 13 principles of green chemistry and engineering for a greener Africa," *Green Chemistry*, vol. 13, 2011, pp. 1059–1060.

17. M. J. Mulvihill et al., "Green chemistry and green engineering: A framework for sustainable technology development," *Annual Review of Environment and Resources*, vol. 36, 2011, pp. 271–293.

18. C. J. Gonzalez et al., "Key green engineering research areas for sustainable manufacturing: A perspective from pharmaceutical and fine chemicals manufacturers," *Organic Process Research & Development*, vol. 15, 2011, pp. 900–911.

19. K. V. Katti, "Realms of green nanotechnology," *International Journal of Green Nanotechnology*, vol. 1, no. 1, 2013.

20. "5 examples of green building in engineering," http://blog.vipstructures.com/5-examples-green-building-engineering/

21. M. N. O. Sadiku, S. R. Nelatury, and S.M. Musa, "Green energy: A primer," *Journal of Scientific and Engineering Research*, vol. 5, no.7, 2018, pp. 336–339.

22. D. Pleissner, "Green chemistry and the leisure industry: New business models for sustainability," *Current Opinion in Green and Sustainable Chemistry*, vol. 8, 2017, pp. 1–4.

23. "The pros and cons of green alternatives in chemical manufacturing," August 2017, https://www.boropharm.com/pros-cons-green-alternatives-chemical-manufacturing/

24. M. Bhardwaj and Neelam, "The advantages and disadvantages of green technology," *Journal of Basic and Applied Engineering Research*, vol. 2, no. 22, October–December, 2015, pp. 1957–1960.

25. K. J. M. Matus et al., "Barriers to the implementation of green chemistry in the United States," *Environmental Science & Technology*, vol. 46, 2012, pp. 10892–10899.

26. A. M. Flynn et al., "Teaching teachers to teach green engineering," *Journal of STEM Education*, vol. 7, no. 3 & 4, July–December 2006, pp. 13–24.

27. D. R. Shonnard et al., "Green education through as U.S. EPA/academia collaboration," *Environmental Science & Technology*, vol. 37, no. 23, 2003, pp. 5453–5462.

28. D. T. Allen et al., "Green engineering education in chemical engineering curricula: A quarter century of progress and prospects for future transformations." *ACS Sustainable Chemistry & Engineering*, vol. 4, 2016, pp. 5850–5854.

29. D. T. Allen and D. R. Shonnard, *Green Engineering: Environmentally Conscious Design of Chemical Processes*. Upper Saddle River, NJ: Prentice Hall, 2002.

30. R. E. Weiner and R. A. Matthews, *Environmental Engineering*, 4th ed. Oxford, UK: Butterworth-Heinemann, 2003.

31. P. T. Anastas and J. C. Warner, *Green Chemistry Theory and Practice*. New York: Oxford University Press, 1998.

32. B. Torok and T. Dransfield, *Green Chemistry: An Inclusive Approach*. Cambridge, MA: Elsevier, 2017.

33. M. A. Ryan and M. Tinnesand, *Introduction to Green Chemistry*. Washington, DC: American Chemical Society, 2002.

34. M. Lancaster, *Green Chemistry: An Introductory Text*, 2nd ed. Cambridge, UK: Royal Society of Chemistry, 2010.

35. A. E. Marteel-Parrish and M. A. Abraham, *Green Chemistry and Engineering: A Pathway to Sustainability*. Hoboken, NJ: John Wiley & Sons, 2014.

36. S. K. Sharma and A. Mudhoo, *Green Chemistry for Environmental Sustainability*. Boca Raton. FL: CRC Press, 2011.

4

Green Revolution

Accept the past for what it was. Acknowledge the present for what it is.
Anticipate the future for what it can become.

—Tracy L. McNair

4.1 Introduction

Food and agriculture are of paramount importance in any society. Food demand is a complex process that depends on the population size, culture, and human behavior such as habits, traditions, and tastes. Although the principal objective of agriculture is to feed people, the sector has always provided nonfood products such as wool, leather, bioenergy, and agrochemicals [1].

Economics scholars have long argued that agriculture plays a critical role in the development of a nation. Agriculture is the largest industry in the world, feeding billions of people. It is regarded as the key to improving livelihoods, food security, and nutrition.

However, the rapid increase in population, adding one billion people every 14 years, places a heavy burden on the agriculture sector to meet the consequential food demand. The traditional agricultural technology was not able to produce enough food for everyone. In order to be able to feed everyone, there was a need to introduce the latest in science and technology to agriculture production. As shown in Figure 4.1, agriculture may be regarded as a technology that mediates between humans and natural resources [2].

Humans have witnessed many revolutions that have dramatically changed lives, such as the American Revolution and the Industrial Revolution. Four of these revolutions are portrayed in Figure 4.2. Another revolution, known as the Green Revolution (GR), occurred and changed the agriculture sector. The GR refers to the transformation in agricultural practices in many parts of the developing nations (such as Mexico, India, Pakistan, Tanzania, Nigeria, Ghana, Malaysia, and the Philippines) that led to a significant increase in agriculture production between 1940 and the 1960s. It occurred during a period when the productivity of global agriculture increased drastically as a result of new advances. It is a combination of controlling chemicals in the soil

FIGURE 4.1

Agriculture mediates between human culture and nature [2].

and pest control and mechanization of agriculture. The revolution sought to replace subsistence agriculture with commercial agriculture.

This chapter provides an introduction to GR, its impacts, advantages, and disadvantages. It also covers the second GR that seeks to correct the limitations of the first GR.

4.2 Concept of Green Revolution

The GR (or the Third Agricultural Revolution) refers to the development of technology transfer initiatives occurring between the 1930s and the late 1960s. The term "Green Revolution" was first used in March 1968 by the former US Agency for International Development (AID) director William S. Gaud. It was funded by the Rockefeller foundation, the Ford foundation, the World Bank, the US AID, aid agencies, and government agencies around the world. Some manufacturers in the United States discovered that it is possible to create a fertilizer from petroleum (the so-called petrochemical fertilizer) that can be used on crops. The GR spread technologies that already existed but had not been widely implemented outside industrialized nations [3]. It was championed by American agronomist Norman Borlaug, who received the Nobel Peace Prize in 1970 in recognition of his efforts. He

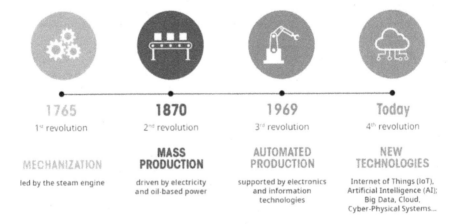

FIGURE 4.2

Four industrial revolutions.

Source: https://www.visiativ-industry.ch/industrie-4-0/

was one of the leading researchers in an international crop improvement program. Dr. Norman Borlaug is usually recognized as the "Father of the Green Revolution."

Today, most farmers practice modern farming under GR. The GR is an alternative solution pushed by the government to replace traditional agricultural means of growing crops.

4.3 Characteristics of Green Revolution

The GR would not have been possible without earlier scientific breakthroughs. It refers to the application of science and technology to increase agricultural production and productivity. It is based on the basic assumption that technology is a superior means of farming. The GR transformed farming practices in many regions of the tropics and subtropics. Some of the major changes in agricultural practices due to the GR are summarized as follows [4, 5]:

- Mechanization on a broad scale, which was a major change
- Implementation of uniform monoculture, which is a characteristic of GR farming
- Package of inputs and modern varieties and high-yielding varieties
- Adoption of new crop varieties
- An increase in the use of machinery and fertilizers
- An increase in irrigational facilities and agricultural credit
- Higher productivity by implementing multiple cropping
- The use of pesticides by farmers to kill pests

Timing is a critical aspect of revolutions. The GR spread technologies that had existed only in the industrialized world. The technologies included irrigation projects, pesticides, synthetic nitrogen fertilizer, and improved crop varieties.

4.4 Green Revolution in Emerging Economies

The GR was a great technological success story of the second half of the 20th century. It thrives in emerging economies including India, Africa, Latin America, China, and much of Asia [6].

- *Mexico:* Some regard 1944 as the beginning of the GR when the Rockefeller Foundation tried to improve Mexican agricultural output. For this reason, some assert that the GR began in Mexico. The results were so astounding that Mexico was transformed from being an importer of wheat to an exporter [7]. The government created the Mexican Agricultural Program (MAP) that was designed to help organizations raise productivity. The GR experience in Mexico has shown that rapidly increasing agricultural production and productivity are feasible in a relatively short time. The success of the program was repeated elsewhere in the developing world. Mexico became the showcase for extending the GR to other areas of Latin America.

- *India:* From its independence in 1947 till 1965, India's agricultural production could not meet the country's needs. Since independence, GR became the catchword in India. The GR was introduced in 1967–1968 in India. To increase agricultural production, the government of India invited a team of experts sponsored by the Ford Foundation. In India, GR comprised three factors: the continual expansion of farming areas, doubling of the existing output of crops, and making use of genetically improved seeds. It completely replaced the traditional farming system. Through GR, India was able to feed vast populations by relying on a genetically engineered rice variant. The GR led to a spectacular increase in agricultural production of food grains, especially the wheat, and brought prosperity to the farmers. The GR met with its greatest success in the Indian state of Punjab. However, Punjab, being the "showcase" of Indian agriculture, is now facing socioeconomic crisis. Practical application of GR strategies has been uneven geographically. The agricultural modernization was experienced differentially by households of different economic status. Large farms benefit more from GR because they can afford canal irrigation, while the small farmers may need loans with high interest rates to irrigate their farms [8, 9]. The GR technology helped to increase yields and prevent famines, but the technology has equally increased inequality, perpetuating the social and economic disparity that already existed. India may need another agricultural technological advancement to keep up with the growing population [10].

- *Africa:* African leaders have acknowledged that agriculture plays a crucial role in their economic development and that lack of investment in the sector would only leave them farther behind. Farmers in Africa are faced with challenges such as unstable governments, widespread corruption, and a lack of infrastructure, particularly less developed roads and water resources. Subsistence farming in Africa depends mainly on family labor, which sometimes requires that children are kept out of school to assist. Agencies like the World Bank stressed fiscal discipline, leading African nations to withdraw support for agriculture. The Rockefeller Foundation along with the Gates Foundation launched a GR in Africa, the Alliance for a Green Revolution in Africa (AGRA). The revolution was designed to transform farming practice and reduce poverty in Africa. This involved a shift from traditional modes of agriculture production. The good news is that the economies of African countries are growing and the governments are supporting the agricultural sector. Political will is being demonstrated through bold actions by a number of African governments [11, 12]. For example, rice is produced in large quantities since it is a highly commercial crop (followed by maize, millet, and sorghum) that plays a key role in providing food security for low-income households. It is debatable whether the GR can deliver Africa from hunger and poverty in the 21st century.

- *China:* China's rapid economic development over the last three decades has catapulted the nation into one of the world's largest economies. Like other nations, China faced the challenge of rapid urbanization and food production. After the end of the Great Leap Forward (GLF) famine in 1961, Chinese political and scientific leaders were faced with producing more food and employing more people on shrinking arable land. By forcing the growing population and workforce to remain diffused in communes, Chinese leaders sought to soften the acute negative effects of the population growth. China's population surge along with the consequent demands on food production was the primary driver of Chinese policymaking after 1970. China's 1970s GR was the result of a nationwide, vertically integrated agricultural extension system. It demonstrates how 1970s China took advantage of high returns to capital to increase food production [13]. China is still a labor surplus economy. Its strategic challenge is to pursue sustainable development, while meeting the expectations of improved living standards and job creation. China is also making great efforts to develop renewable energy in order to cut down on emissions. China's new industrial strategy will prioritize the development of eight industries: alternative energy, biotechnology, new-generation information technology, high-end equipment manufacturing, advanced materials, alternative-fuel cars, energy saving, and environmental protection [14].

The same success story can be told of Brazil, Vietnam, Pakistan, the Philippines, Malaysia, and Korea.

4.5 Impacts of the Green Revolution

From a global perspective, the GR is regarded as an extraordinary success.

The GR has had a great impact on food production, socioeconomic conditions, politics, environmental sustainability, and technology [15, 16].

- *Impact on food production:* The GR has led to exponentially increased food production worldwide, particularly in the developing world. For example, total rice production in Asia increased by 132% from

1966 to 1999. During the same period, world wheat production increased by 91%, to 576 million tons. The GR has transformed India into a food-surplus country. Food grain production has increased significantly after introduction of the high-yielding varieties of crops (HYV).

- *Socioeconomic impact:* The socioeconomic impacts of the GR are mixed. The decline in food prices has benefited both the urban poor and the rural landless. The GR contributed to better nutrition by raising incomes and reducing prices. The GR also created plenty of jobs for both agricultural workers and industrial workers.

- *Political impact:* Critics of the GR claim that the primary objective of the program was geopolitical: to provide food for the populace in undeveloped countries and weaken socialist movements in many nations.

- *Environmental impact:* The availability of cereal varieties with multiple resistance to diseases has reduced agrochemical use, thereby improving the health of farming communities. The GR relies on heavy use of chemical inputs. This has resulted in a precarious situation of water scarcity and vulnerability to pests. The water quality is being compromised and rivers are drying up at a rapid rate.

- *Technological impact:* The GR relied on technology to create a new mode of agricultural production. Technologically, the GR introduced an agricultural mode that followed industrial logic. The use of fertilizers, modern machinery, and extensive irrigation facilities also contributed to significant increases in agricultural output.

Based on these impacts, the GR is considered as successful. Some studies claim a miraculous transition to high productivity growth rates. The rapid productivity and production gains have served a crucial purpose in helping achieve global food security.

4.6 Advantages

The GR has been a key ingredient for initiating and supporting economic transformation in many countries. To have a well-informed decision, it is best to look at the main advantages and disadvantages of GR. The major advantages include the following [17]:

1. It allows agricultural operations on a massive scale
2. It makes plants that are resistant to pests and herbicides
3. It has the potential to be able to grow any crop anywhere

4. It eliminates the need to fallow lands

5. It has led to a great increase in the production of food grains (especially wheat and rice)

6. It has positive impacts on poverty reduction and lower food prices

7. It allows farmers to grow and export cash crops for profit

8. It creates job opportunities in the agricultural and industrial sector

9. It increases agricultural production and productivity

10. It decreases amount of human labor

11. It allows more food on the same amount of land

12. It allows automation in the process of farming

An important effect of the GR is that traditional agricultural practices have been replaced by scientific practices. The GR has caused increased agricultural production by overcoming cultural and religion constraints on technology. Although increased food production should benefit both farmers and consumers, this is not always the case. Increased supply can cause rapid declines in food prices, thereby causing revenue losses for farmers and discouraging production [18]. Some of the benefits of GR are illustrated in Figure 4.3 [19].

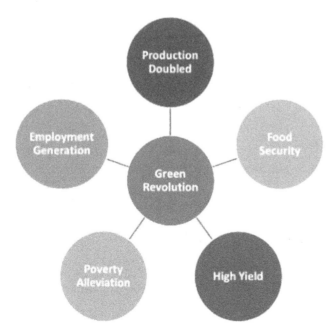

FIGURE 4.3
Some of the benefits of the Green Revolution [19].

4.7 Disadvantages

The critics of the GR insist that GR is an agricultural model that relied on a few staple and market profitable crops. They claim that it also produces problems.

Problems or disadvantages of the GR include the following [17]:

1. It causes pests and weeds to develop hazards
2. It employs monoculturing
3. It results in both depletion and pollution of water
4. It tends to reduce the natural fertility of the soil
5. It requires new machines, which means less people are needed to farm
6. The cost of farming equipment is expensive for small farmers
7. Income inequalities increased between rich and poor farmers
8. Pesticides and herbicides are a health hazard to farmers
9. Women farmers have gained proportionally less than their male counterparts
10. Undesirable social consequences

A revolution of this magnitude created some problems of its own. GR has been widely criticized for causing environmental damage and increased income inequality. Few nations can afford the equipment that GR programs require. GR brought more income to the already rich, thereby widening the gap between the rich and the poor.

4.8 Second Revolution

To combat climate change and hunger, a number of governments, foundations, and aid agencies have called for a "New Green Revolution," with more integrative environmental and social impacts combined with agricultural and economic development [20]. The degrees of freedom for sustainable human exploitation of the planet Earth are severely restrained. Global freshwater resources are a fundamental limiting factor in feeding the world. This raises the urgency of developing strategies to build resilience in water resource management [21].

Although the first GR helped alleviate food shortage in several developing nations, a central contention has been that the GR overlooked the rural poor and tended to favor the minority of commercial farmers. The GR has also resulted in both depletion and pollution of water. Although the GR has improved agricultural output in some regions in the world, there is still room for improvement. During the post-GR period, there has been renewed interest in agricultural investment, and there are calls for the next GR. There has been a recognition of the limitations of the first GR and the need to correct these limitations and minimize unintended negative consequences. Due to this, the second GR (GR 2.0) will likely focus on improving tolerances to pests and disease in addition to technological input use efficiency. Population growth, ongoing exploitation of natural resources, and increasing costs of chemical fertilizer will make GR 2.0 a priority. GR 2.0 is already taking place in low-income nations [22]. GR 2.0 must focus on shifting the yield frontier for the major staples. It should be all inclusive in its coverage of small farmers, rainfed areas, sustained use of resources, and application of nanotechnology.

4.9 Conclusion

The human need for food creates a vital relationship between us and the environment. GR is the agricultural transformation of less developed nations. It has been a milestone in the international agricultural movement and has had an enormous impact. It has not been a red, bloody revolution as critics predicted. Rather, it has been heralded as a political and technological achievement that is unprecedented in human history. The GR along with the development of new technologies, new business models, and evolving consumer demands has drastically affected the global agri-food system. Achieving a GR requires an increase in public investments in agriculture, rural infrastructure, and marketing.

The GR was successful for a number of reasons. There has been increased food production commensurate with the growing population. There has been a close relationship between farmers, scientists, and policy makers. Some have hailed GR as the greatest thing that has happened to the developing world. More information about the GR can be found in books in References [1, 23–30] and in other books available on Amazon.com.

References

1. J. I. Boye and Y. Arcand (eds.), *Green Technologies in Food Production and Processing.* Boston, MA: Springer, 2012.
2. J. H. Perkins, *Geopolitics and the Green Revolution: Wheats, Genes, and the Cold War.* New York: Oxford University Press, 1997, p. 5.
3. "Green revolution." *Wikipedia*, the free encyclopedia https://en.wikipedia.org/wiki/Green_Revolution

4. A. Goldman and J. Smith, "Agricultural transformations in India and Northern Nigeria: Exploring the nature of Green Revolutions," *World Development*, vol. 23, no. 2, 1995, pp. 243–263.

5. "Causes or Importance of green revolution (GR)," http://www.economicsconcepts. com/causes_or_importance_of_green_revolution_(gr).htm

6. M. N. O. Sadiku, S. R. Nelatury, and S. M. Musa, "A primer on green revolution," *Journal of Scientific and Engineering Research*, vol. 5, no. 7, 2018, pp. 332–335.

7. A. Ameen and S. Raza, "Green revolution: A review," *International Journal of Advances in Scientific Research*, vol. 3, no. 2, 2017, pp. 129–137.

8. K. Sebby, "The green revolution of the 1960's and its impact on small farmers in India," *Undergraduate Thesis*, University of Nebraska-Lincoln, January 2010.

9. S. Dutta, "Green Revolution revisited: The contemporary agrarian situation in Punjab, India," *Social Change*, vol. 42, no. 2, 2012, pp. 229–247.

10. A. Panning and K. G. Kulkarni, "Technology for growth: Indian green revolution," *SCMS Journal of Indian Management*, vol. VIII, July–September 2011, pp. 49–62.

11. P. A. Sanchez, G. L. Denning, and G. Nziguheba, "The African green revolution moves forward," *Food Security*, vol. 1, 2009, pp. 37–44.

12. C. Breisinger et al., "Potential impacts of a green revolution in Africa—The case of Ghana," *Journal of International Development*, vol. 23, 2011, pp. 82–102.

13. J. Eisenman, "Building China's 1970s green revolution: Responding to population growth, decreasing arable land, and capital depreciation," https://www. researchgate.net/publication/319195398_Building_China's_1970s_Green_ Revolution_Responding_to_Population_Growth_Decreasing_Arable_Land_ and_Capital_Depreciation

14. "China's green revolution energy, environment and the 12th five-year plan," http://www.greengrowthknowledge.org/sites/default/files/downloads/ resource/China%E2%80%99s_Green_Revolution_Energy_Environment_and_ the_12th_Five-Year_Plan_Chinadialogue.pdf

15. G. S. Khush, "Green revolution: The way forward," *Nature Reviews Genetics*, vol. 2, 2001, pp. 815–822.

16. R. E. Evenson, "The green revolution in developing countries: An economist's assessment," http://unpan4.un.org/intradoc/groups/public/documents/APCITY/ UNPAN020402.pdf

17. "6 Advantages and disadvantages of the green revolution," https://futureof-working.com/6-advantages-and-disadvantages-of-the-green-revolution/

18. X. Diao, D. Headey, and M. Johnson, "Toward a green revolution in Africa: What would it achieve, and what would it require?" *Agricultural Economics*, vol. 39, 2008, pp. 539–550.

19. "Green revolution in India," https://iasmania.com/green-revolution-in-india/

20. R. Patel, "The long green revolution," *The Journal of Peasant Studies*, vol. 40, no. 1, 2013, pp. 1–63.

21. J. Rockstrom and L. Karlber, "The quadruple squeeze: Defining the safe operating space for freshwater use to achieve a triply green revolution in the Anthropocene," *AMBIO*, vol. 39, 2010, pp. 257–265.

22. P. L. Pingali, "Green revolution: Impacts, limits, and the path ahead," *Proceedings of the National Academy of Sciences*, vol. 109, no. 31, July 2012, pp. 12302–12308.

23. G. B. Marini-Bettòlo (ed.), *Towards a Second Green Revolution*. Amsterdam, The Netherlands: Elsevier Science, May 1988.

24. M. S. Randhawa, *Green Revolution*. New York: John Wiley & Sons, 1974.

25. G. R. Conway and E. B. Barbier, *After the Green Revolution: Sustainable Agriculture for Development*. Sterling, VA: Earthscan, 1990.

26. P. B. R. Hazell and C. Ramasamy, *The Green Revolution Reconsidered: The Impact of High-yielding Rice varies in South India*. Baltimore, MD: Johns Hopkins University Press, 1994.

27. B. Sen, *The Green Revolution in India: A Perspective*. New Delhi, India: Wiley Eastern Private Ltd., 1974.

28. G. Conway, *The Doubly Green Revolution: Food for All in the Twenty-First Century*. Ithaca, NY: Cornell University Press, 1997.

29. V. Shiva, *The Violence of the Green Revolution: Third World Agriculture, Ecology and Politics*. Lexington: University Press of Kentucky, 2016.

30. D. Brautigam, *Chinese Aid and African Development: Exporting Green Revolution*. Basingstroke: Macmillan, 1998.

5

Green Agriculture and Food

> The recipe for success is to study while others are sleeping, work while others are loafing, prepare while others are playing, and dream while others are wishing.
>
> —**William A. Ward**

5.1 Introduction

Agriculture has made an enormous environmental footprint and is making agricultural development risky. Food is an integral part of human existence. Industrialization, climate change, and rising population have made evident the precarious balance between sustainable food production practices, a healthy environment, and healthy population. Our daily food and drink consumption makes up a noticeable proportion of global greenhouse gas emissions. Rising demand means we have to produce and consume in more efficient and less damaging ways [1].

Green agriculture uses well-developed modern farming and sustainability concepts to improve natural agricultural techniques. It also draws on green technology to enhance farming. "Green" foods have emerged as alternatives for consumers wishing to avoid "conventional" foods produced using chemicals.

Green products in general are known as environment-friendly products. Green foods refer to foods that are produced under the principle of sustainable development. They are safe for consumption; they are of fine quality, nutritious, healthy, safe, and environmentally friendly [2]. Green food with less chemical residuals has become more popular around the globe. Governments and activists have attempted to address environmental and health problems in the food industry through promoting green food. As illustrated in Figure 5.1, there is a close relationship between food, health, economy, security, and the environment [3].

This chapter introduces the concepts of green agriculture and green food. It begins with a discussion on green agriculture, its concept, applications, benefits, and challenges. It addresses the three types of foods and explains the concept of green food. It covers constructing green food and green food consumption. The chapter also discusses the benefits and challenges of green food. The last section concludes the chapter.

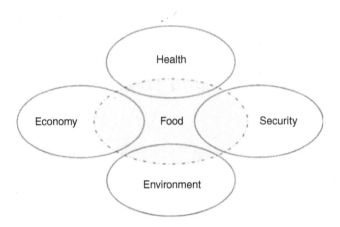

FIGURE 5.1
The relationship between food, health, economy, security, and the environment [3].

5.2 Green Agriculture

Modern agriculture has contributed substantially to environmental and health issues. For example, agriculture is responsible for 70% of water withdrawals worldwide and causes water pollution. Increasing threats of climate change and agricultural pollution have influenced public officials to think proactively about ways to produce and deliver food more efficiently. Inefficient or unsustainable agriculture may eventually exhaust the available resources.

Green or sustainable agriculture is basically farming in sustainable ways. Unlike traditional agriculture, where the profit margin is the single major factor, green agriculture is defined by three main factors: economic profit, environmental stewardship, and social responsibility.

Sustainability of natural resource management systems is defined by seven general attributes: productivity, stability, reliability, resilience, adaptability, equity, and self-reliance [4]. Green agriculture uses adaptable local farming techniques and practices that increase farming yields. Renewable sources of energy for agriculture are a green option for cutting back on energy consumption. Solar, thermal, photovoltaic, geothermal, wind, and water power are possible options.

5.2.1 Concept of Green Agriculture

The concept of green agriculture stems from the green economy principles that state that economy-driven growth must be more resource-efficient and cleaner. Green economy is one that results in increased human well-being, while significantly reducing environmental risks.

The UN has provided the following five key principles of green agriculture [5]:

- Integration of livestock crops
- Use of postharvest storage and processing facilities to reduce waste
- Ensuring that crop rotations are diversified
- Use of environmentally sustainable weed and pest control practices
- Use of natural and sustainably made nutrient inputs

The UN mentions the following six practices that must be improved in order to transition from traditional agriculture to green agriculture [5]:

- Diversifying crops and livestock
- Making water use efficient and thus more sustainable
- Managing soil fertility
- Ensuring that agriculture storage facilities are more efficient and sustainable
- Improving the management of animal and plant health
- Introducing new and environmentally friendly and labor-friendly modes of mechanization

There is no single approach to adopting green agriculture; the precise goals and methods must be adapted to each individual case. There are several ways to achieve a sustainable agriculture that provides enough food and ecosystem services for present and future generations. Not all geographic regions lend themselves readily to green agriculture. Greenhouses have also been used to provide green agriculture to arid climates.

5.2.2 Applications

Greening of agriculture is far more than a simple study of the marketplace and expanding the advertising process. Green agriculture affects many economic sectors. The applications that are considered here are only typical. Each application pursues triple bottom line—economics, environmental agenda, social decisions—and impacts the greening of agriculture and the food system.

- *Green agriculture products supply chain:* This chain makes the concept of "green, health, and environmental protection" throughout the agriculture products supply chain. The green agriculture products supply chain management can be divided into green purchasing, green production, green manufacturing, green logistics, green marketing, green consumption, and green recycling. All links are connected into a whole [6].

- *Green jobs*: Agriculture has always been a productive economic sector and will continue to be in future times. Agriculture is labor intensive and specific measures shall be put in place, such as insurance coverage for natural disasters. Green agriculture and green jobs play a pivotal role in fostering social development and sustainability. The transition to a more sustainable agriculture delivers more employment opportunities such as green jobs [7].

- *Organic farming*: Organic farming is the type of farming in which crops develop from natural resources having the complete nutritive value and preventing the crop/plants from the pests. It is a sustainable approach to food production system, an alternative to ecologically unsound practices of conventional farming. Those who started organic farming considered themselves as improving the environment, producing safe and healthy food, and generally filling an important role in the society. With strict rules that govern organic farming, such as prohibition of application of chemical pesticides and fertilizers, most organic farms are greener than most conventional farms [8]. Consumers are increasingly interested in fully organic food and they are willing to pay higher prices for it than for conventionally produced food.

- *Nanotechnology*: This is a new rapidly evolving field with applications covering various areas such as medicine and agriculture. Aspects related to productivity, soil health, water security, and food quality maintenance during storage and distribution can be strongly impacted through nanotechnological interventions. Successful implementation of nanotechnological interventions requires a judicious mix of science, policy, and technologies. In agriculture, nanotechnology can help the agro-food sector in sustainable production as well as in reducing the environmental impact of agricultural operations. Nanoparticles play a major role in plant growth and productivity. However, the application of nanoparticles in agriculture is still in the early stages [9].

5.2.3 Benefits and Challenges

Throughout the world, evidence is mounting to support sustainable forms of agricultural production as viable alternatives to conventional farming practices. Adopting green agriculture results in many localized benefits. Green agriculture products with no pollution, high quality, and high nutritional value are more and more appreciated by the consumers. Green agriculture enhances the well-being of rural populations and poverty reduction in developing countries.

Some major challenges to the sustainability of the world's agriculture are [10]: (i) pollution, (ii) biodiversity loss, (iii) soil degradation/nutrient loss/

erosion, (iv) water scarcity/salinity, (v) carbon footprint, and (vi) natural resource depletion. Simply making a technology available does not mean it will be adopted. Barriers to adopting new technologies can limit their effectiveness.

5.3 Types of Food

Food is the primary necessity of human beings. The food industry has five core sectors [11]: (1) food production (farm), (2) food processing (food, meat, poultry, factories), (3) food distribution (warehouses, trucking), (4) food retail (supermarkets, grocery stores), and (5) food service (restaurants, cafeterias).

There are three kinds of food in terms of environmental friendliness [12]: normal, green, and organic. Normal food has measurable standards of quality and sanitation. Green food implies planting under zero environmental pollution conditions. Organic food develops slowly due to its unclear market positioning. Organically produced food is regarded as healthier, safer, better tasting, and more nutritious than conventionally produced food. The purpose of developing organic food is to promote sustainable agricultural development.

Organic foods are regarded as "credence goods" because some of the attributes that consumers may consider are not obvious or easily verified. Since organic foods are generally more expensive than their conventional ones, only the wealthy or environment conscious could afford them.

Sometimes, the terms green food and organic food are used interchangeably, but the two are not the same. Green food is a "middle way" between chemical and organic farming. Green foods are of two types: green foods that allow the use of some chemicals and organic foods. Thus, all green foods are not organic foods.

Food products, such as fruit and vegetables, fat and oils, sugar, dairy, meat, coffee, and flours are complex mixtures of vitamins, sugars, proteins, fibers, and other organic compounds. Before such products can be commercialized, they have to be processed and preserved for food-ready meals [13].

5.4 Concept of Green Food

The green civilization is promoting the green economy. Food has entered into the rank of green. Green food has become the principal demand of people in developed nations. It is becoming the mark of their living standard. The consumption of green food is growing and has become the international trend [14]. America has become the largest green (or organic) food market in the world. The green food industry has been developing rapidly in developed nations such as the United States of America, the United Kingdom, Germany, Denmark, Switzerland, and Austria. It is picking up in the developing and less developed countries.

The concept of "green consumption" was introduced by the International Organization of Consumer Unions in 1963 [15]. Since then sustainable farming awareness has increased. The concept of green food was born in response to some needs including the Green Revolution (see Chapter 1), environmental pollution, pesticide, contaminations, food insecurity, climate change, biodiversity, and water availability. The term "green" figuratively represents the properties of pollution-free, safety, and healthiness. Providing food that is sustainable will be one of the most important challenges facing the agriculture sector and the food industry. This is not something that can be achieved by just the agriculture and food sectors.

A realistic strategy toward sustainable food growth requires the cooperation of the entire food chain. Sustainable agro-food systems that produce sustainable food must embody attributes such as respect for environmental limits, high standards of animal welfare, affordable food, and a viable livelihood for farmers. Government has a major role to play in sustainable food issues, providing leadership and setting priorities. It must take bold steps to encourage the provision of a healthier and more sustainable set of choices. To promote the development of green food, regulatory authorities must create a green food environment worthy of the public trust.

5.5 Green Food Construction

Green foods are construed as fresh, chemical-free, nutritious, natural, and are produced in an environmentally sustainable manner. A green-growth strategy for the food and agriculture sector encompasses agriculture, fisheries, and the food supply chain.

Food processing is a set of methods and techniques used to transform raw ingredients into finished and semifinished products. Today, almost all foods are processed in some way. Common conventional food processing techniques include:

- *Heating:* This is one of the most important techniques for food processing. Heating is used to extend life of foods by destroying enzymes, microorganisms, insects, and parasites and removing water to prevent deterioration. It changes the nutritional and sensory qualities of foods. Many foods are consumed in a cooked form that has flavors that cannot be created otherwise. This may involve cooking, baking, frying, and roasting. Thermal processing methods such as boiling, frying, and roasting are frequently applied to certain foods prior to consumption. Pressure cooking and microwave heating represent additional food processing methods.

- *Mixing:* This is a common food processing technique used to evenly distribute each ingredient during manufacturing. Mixing is often required to achieve uniformity in the raw material. An example is mixing of cookie or bread dough.
- *Fermentation:* This has been a traditional part of the diet for centuries. Today, it is used in the food processing industry for making baked products, alcoholic beverages, yoghurt, cheese, soy products, etc.
- *Packaging:* Processed foods are generally available in packages. Packaging serves a number of purposes. First, it plays important roles in preservation and food safety. Second, it can also make food easier to transport, store, and serve. Third, it provides a surface for displaying labels. Fourth, it provides protection against damage, environmental contamination, and tampering. Discarding packaging materials has the potential for pollution. Recycling and reusing containers is environmentally preferred.

Other methods of food processing include dehydration (or drying), frying, chilling, freezing, coating, irradiation, blanching, pasteurization, heat sterilization, and extrusion.

Green food processing techniques include preservation, transformation, and extraction. Different methods can be used for this purpose; for example, frying, drying, filtering, and cooking. Some food products are known to be thermally sensitive and vulnerable to chemical, physical, and biological changes [13].

One way to reduce environmental footprint in food processing is the use of enzymes, which speed up reaction rates. Enzymes in foods may produce extended shelf life, improved textures, functionality, and yield. Drying is used for processing of many bulk and packaged food products and ingredients. Some plants may need the supply of sufficient fat-soluble vitamins for long continued and vigorous growth [16].

Some have suggested that the best way to improve the sustainability of human food is to consume less and reduce consumption of processed foods. Highly processed food is easy to overeat. In restaurants, foods are often served in large portion containers and are intended to be totally consumed [17].

Merely looking at the food does not give the consumer any idea of how it was processed. Having "green food" labels on products will help consumers understand the products and make responsible purchasing decisions. This can accelerate consumers' decisions to buy green food. The marketer of green-labeled food can put some useful information about products on the label such as nutrition level and ingredient [18].

5.6 Green Food Production

Agriculture and food production are integral part of everyday experience for many in developed and developing nations. As American society has transformed from an agricultural to a postindustrial economy, food producers have been somewhat replaced by food consumers. Mass-produced food has become more popular in the vast majority of households. Food production activities are grounded in social relations. The factors that impact food production include global climate changes, safety, security, and ethics.

- *Climate change:* As the world population continues to increase, agricultural productivity will be tasked to keep pace without taxing environmental resources. Maize is one of the most important cereals for food security in some tropical parts of the world, especially in Africa, Asia, and Latin America. Climate change and changes in rainfall distribution will influence production of maize in the tropics.

- *Food safety:* Food safety is the joint responsibility of everyone involved in the food supply chain from farm to the consumers. This is a public health issue that has become a global problem. Since food is necessary for human survival, food safety crisis can cause social panic and heavy casualties. Food safety has been a major concern for government, the food service industry, and academia. The food service industry aims to reduce the risk and occurrence of foodborne illness. Use of mechanized equipment introduces safety issues [19].

- *Food security:* Food security is a fundamental human right. It is defined as year-round access to an adequate supply of safe and nutritious food. The components of food security include availability, access, utilization, and stability. These four pillars must be met to ensure food security. Water supply directly affects food production and ensures food security. Having sufficient water ensures crop growth and livestock survival [20].

- *Food ethics:* Ethics deals with reasoning and judging about rights and wrongs in relations. Food ethics is related to agriculture, its production, marketing, distribution, preparation, and consumption. Food ethics is the ethical thinking in regard to food. The ethical dimensions of food include food security, food safety, and food law.

Green food production is part of green agriculture. It refers to agricultural products that are produced using organic methods of agriculture [21]. Producing green food meets the growing needs to save our environment from destruction. Green food production often evokes organic farming practices. Green foods refer to foods that are high-quality, safe, and healthy to be consumed. For the production of green food, limited amounts of pesticides and synthetic fertilizers are allowed.

5.7 Green Food Consumption

Green food consumption can guarantee high quality of life for consumers and promote green food production. Interest in food products with eco-friendly characteristics is growing. People are "going green" with their spending habits, behaviors, and diets. Green products become more acceptable when they are affordable, accessible, of high quality, and eco-friendly. For example, the restaurant industry is an energy-intensive service industry that requires a great deal of electricity, water, fuel, and agricultural products. Several restaurant managers are willing to implement green features such as using green foods, designing environmental management, and even building green buildings [22].

Consumers' lifestyle, age, religion, health consciousness, green food awareness, income level, price, and shopping convenience are the major factors that influence consumer's behavior of purchasing green food. Consumers' age, education, and income significantly affect green food awareness. Education influences green food consumption and family income dictates the development of green food consumption behavior [23]. Customer value and customer satisfaction significantly affect the customers' loyalty toward green food products.

Green foods can revolutionize our health and happiness. Green foods are packed with nutrients, fiber, and water; they also have great healing. Some popular green foods include green beans, spinach, avocado, green pepper, and broccoli, to name a few. Green beans have a number of health benefits when eaten raw or cooked. Spinach is a versatile food that can be prepared in a variety of tasty ways. It has a lot of health benefits. Avocado is loaded with healthy fat that helps lower bad cholesterol. Green pepper is very low in calories and is known for its ability to promote digestive health. Adopting and incorporating these foods into your regular diet will do wonders for your health [24].

In addition, the following eight principles for healthy and sustainable diets are helpful [25]:

1. Eat a varied balanced diet to maintain a healthy body weight.
2. Eat more plant-based foods, including at least five portions of fruits and vegetables per day.
3. Value your food. Ask about where it comes from and how it is produced. Don't waste it.

4. Moderate your meat consumption, and enjoy more peas, beans, nuts, and other sources of protein.

5. Choose fish sourced from sustainable stocks.

6. Include milk and dairy products in your diet or seek out plant-based alternatives, including those that are fortified with additional vitamins and minerals.

7. Drink tap water.

8. Avoid foods high in fat, sugar, and salt.

5.8 Benefits and Challenges

Some benefits of green foods are illustrated in Figure 5.2 [26]. Green food is a type of food with pollution-free, safe, high-quality, and nutritious content. It is healthy to consume, uses less chemicals, and has higher vitamin and mineral content than conventional foods. The basic objectives of developing green food are to enhance food quality, to promote consumers' health, and to protect ecological environments for sustainable development [27]. Green food consumption can guarantee the life quality of consumers. It can promote green food production. The green food company stands to benefit from the increase in demand for green food.

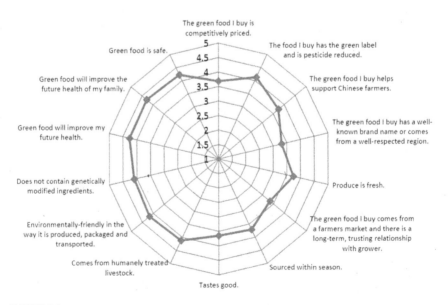

FIGURE 5.2
The benefits of green food [26].

The wide variety of foods available to consumers poses a huge challenge for eating less, improving sustainability, and measuring the carbon impact of a meal. There are challenges associated with the carbon footprint of food items. The primary challenge is that food production is inherently dependent on nature. Another challenge is related to the emission intensity of the production process and the supply chain. The green revolution was confined to certain crops (rice, wheat) at the expense of others. Culture often dictates to some extent how much can change. There are problems associated with traditional marketing of green foods, but web marketing has been effective. The high prices of green foods and the need for more market penetration are a dilemma. For green food, safety remains a great concern. Green food processing techniques need standardization. For each nation, government has a key role to play in sustainable food issues by providing leadership and indicating priority areas for action. The government also creates opportunities for "green jobs" by encouraging and investing in the green food industry [28].

5.9 Conclusion

Agriculture affects the natural environment in providing for management of land and water resources. Green agriculture incorporates profit, environmental stewardship, fairness, health, business, and familial aspects on a farm setting. Using green technologies and farming practices will allow us to continue with the tradition of agriculture without harming the planet.

The concept of green foods is developing in developed nations, but it is still at its infancy stage in developing nations. Although green food consumption is a global trend, it is not an easy task. Environmental value of consumers is necessary to motivate green food consumption, since this influences customer satisfaction of green food products [29]. With growing awareness green food will be the mainstream on table and will improve the health of our families in the future. More information on green agriculture and green food can be found in books in 'References' [30–34] and other books on the subject available on Amazon.com.

References

1. M. N. O. Sadiku, S. M. Musa, and O. S. Musa, "Green food," *Invention Journal of Research Technology in Engineering and Management*, vol. 2, no. 7, July 2018, pp. 20–22.
2. M. R. T. Khan, S. Chamhuri, and H. S. Farah, "Green food consumption in Malaysia: A review of consumers' buying motives," *International Food Research Journal*, vol. 22, no. 1, 2015, pp. 131–138.

3. J. I. Boye and Y. Arcand, "Current trends in green technologies in food production and processing," *Food Engineering Reviews,* vol. 5, no. 1, 2013, pp. 1–17.

4. P. Koohafkan, M. A. Altieri, and E. H. Gimenez, "Green Agriculture: foundations for biodiverse, resilient and productive agricultural systems," *International Journal of Agricultural Sustainability,* vol. 10, no. 1, 2012, pp. 61–75.

5. "What is green agriculture?" http://www.greencityplumber.ca/blog/going-green-tips/green-agriculture/

6. B. Zhao and J. Liu, "Research on the model construction of modern green agriculture products supply chain," *IEEE International Symposium on IT in Medicine and Education,* December 2011, pp. 173–176.

7. A. Bianco, "Green jobs and policy measures for a sustainable agriculture," *Agriculture and Agricultural Science Procedia,* vol. 8, 2016, pp. 346–352.

8. C. Francis et al., "Greening of agriculture," *Journal of Crop Improvement,* vol. 19, no. 1–2, 2007, pp. 193–220.

9. H. Mahawar and R. Prasann. "Prospecting the interactions of nanoparticles with beneficial microorganisms for developing green technologies for agriculture," *Environmental Nanotechnology, Monitoring & Management,* vol. 10, 2018, pp. 477–485.

10. A. Hall and K. Dorai, "The greening of agriculture: Agricultural innovation and sustainable growth," Unknown Source.

11. M. L. Clayton et al., "U.S. food system working conditions as an issue of food safety: Key stakeholder perspectives for setting the policy agenda," *New Solutions: A Journal of Environmental and Occupational Health Policy,* vol. 26, no. 4, 2017, pp. 599–621.

12. G. E. Bekele, "Analysis of organic and green food production and consumption trends in China," *American Journal of Theoretical and Applied Business,* vol. 3, no. 4, 2017, pp. 64–70.

13. F. Chemat et al., "Review of green food processing techniques: Preservation, transformation, and extraction," *Innovative Food Science and Emerging Technologies,* vol. 41, 2017, pp. 357–377.

14. Y. Zhao, "Heilongjiang's green food trade strategies towards Russia," *Proceedings of International Conference on Management Science & Engineering,* Harbin, China, August 2007. pp. 1428–1433.

15. A. Leggett, "Bringing green food to the Chinese table: How civil society actors are changing consumer culture in China," *Journal of Consumer Culture,* 2017, pp. 1–19.

16. T. B. Osborne and L. B. Mendel, "The vitamins in green food," *Journal of Biology & Chemistry,* vol. 37, 1917, pp. 187–200.

17. M. C. Schraefel, "Green food through green food: A human centered design approach to green food technology," *Proceedings of the 2013 ACM Conference on Pervasive and Ubiquitous Computing,* September 2013, pp. 595–598.

18. Z. Bing, S. Chaipoopirutana, and H. Combs, "Green product consumer buyer behavior in China," *American Journal of Business Research,* vol. 4, no.1, 2011, pp. 55–71.

19. M. N. O. Sadiku, S. M. Musa, and O. S. Musa, "Food Safety: A primer," *International Journal of Research Technology in Engineering and Management,* vol. 2, no. 6, June 2018, pp. 1–4.

20. M. N. O. Sadiku, S. M. Musa, and O. S. Musa, "Food Security: A primer," *Invention Journal of Research Technology in Engineering and Management,* vol. 2, no. 7, July 2018, pp. 16–19.

21. J. Adams and J. Wang, "Industry insights Industrial clusters and regional economic development in China: The case of 'green' food," *Journal of Chinese Entrepreneurship*, vol. 1, no. 3, 2009, pp. 279–294.

22. Y. F. Wang, "Development and validation of the green food and beverage literacy scale," *Asia Pacific Journal of Tourism Research*, vol. 21, no. 1, 2016, pp. 20–56.

23. Z. Dawei and W. Mengdi, "Research on the influencing factors of consumer purchase behavior of green food in Harbin," *Proceedings of the 28th Chinese Control and Decision Conference*, 2016, pp. 4479–4482.

24. E. Vyverberg, "These 6 green foods are the key to a healthy and happy lifestyle," https://www.hartigdrug.com/blog/green-foods

25. "Sustainable consumption report: Follow-up to the green food project," July 2013, https://www.gov.uk/government/publications/sustainable-consumption-report-follow-up-to-the-green-food-project

26. B. L. McCarthy, H. B. Liu, and T. Chen, "Trends in organic and green food consumption in China: Opportunities and challenges for regional Australian exporters," *Journal of Economic and Social Policy*, vol. 17, no. 1, 2015.

27. L. Lin, D. Zhou, and M. Caixue, "Green food industry in China: Development, problems and policies," *Renewable Agriculture and Food Systems*, vol. 25, no. 1, 2009, pp. 69–80.

28. M. N. O. Sadiku, A. A. Omotoso, and S. M. Musa, "Green agriculture," *International Journal of Trend in Scientific Research and Development*, vol. 4, no. 2, February 2020.

29. Q. Zhu et al., "Green food consumption intention, behaviors and influencing factors among Chinese consumers," *Food Quality and Preference*, vol. 28, 2013, pp. 279–286.

30. D. Sandoval, *The Green Foods Bible*. Panacea Publishing, 2007.

31. G. Cousens, *Rainbow Green Live-Food Cuisine*. Berkeley, CA: North Atlantic Books, 2003.

32. J. Boye and Y. Arcand (eds.), *Green Technologies in Food Production and Processing*. New York: Springer, 2012.

33. P. B. Thompson and D. M. Kaplan (eds.), *Encyclopedia of Food and Agricultural Ethics*. Netherlands: Springer, 2014.

34. A. Kahn, *Green Agriculture Newer Technologies*. Jaipur, India: Agrotech Press, 2014.

6

Green Nanotechnology

> Commit yourself to lifelong learning. The most valuable asset you will ever have is your mind and what you put into it.
>
> —Brian Tracy

6.1 Introduction

Nanotechnology refers to the science of nanomaterials. It is the measuring, modeling, and manipulating of matter at atomic scale or in the dimension of 1–100 nanometers (nm). (A nanometer is 1 billionth of a meter.) It offers the opportunity to produce new structures, materials, and devices with unique properties such as conductivity, strength, and chemical reactivity. Electrical and mechanical properties can change at the nanoscale. For example, at the nanoscale, gold becomes an active catalyst, helping to turn chemicals X and Y into product Z. Most nanomaterials are made by chemical processes that may or may not generate pollutants or waste materials. There is a social responsibility concerning the production and use of nanomaterials and their potential environmental deleterious effects. Green nanotechnology (GN) is a sustainable approach to nanotechnology.

GN is the use of products of nanotechnology to enhance sustainability. It involves making green nano products that are more environmentally friendly throughout their life cycle and using them to support sustainability. GN has important implications for medical biology, energy shortages, materials science, and a number of emerging disciplines.

This chapter introduces the readers to GN. It begins by introducing nanotechnology. It discusses the main features of GN and its principles. It presents some applications of GN. It highlights some of the benefits and challenges of GN. The last section concludes the chapter.

6.2 Nanotechnology

Nanotechnology (science on the scale of single atoms and molecules) has been called the second Industrial Revolution because of the special properties of materials at the nanoscale. It is a branch of green technology that has the potential to revolutionize many aspects of our lives. It has permeated all sectors of our economy due to the unique properties of materials at the nanoscale. It is transforming the world of materials and its influence will be broad. It will not only initiate the next industrial revolution, it will offer technological solutions.

The term "nanotechnology" was coined in 1974 by Norio Tanigutchi, a professor at Tokyo Science University. Nanotechnology is the science of small things—at the atomic or nanoscale level. It has the idea that the technology of the future will be built on atoms. It has impact on every area of science and technology [1]. Nanotechnology involves imaging, measuring, modeling, and manipulating matter at the nanoscale. At this level, the physical, chemical, and biological properties of materials fundamentally differ from the properties of individual atoms and molecules or bulk matter [2].

Nanotechnology covers a wide variety of disciplines like physics, chemistry, biology, biotechnology, information technology, engineering, and their potential applications. Some of the sectors covered by nanotechnology are shown in Figure 6.1 [3]. Nanotechnology holds great potential for pollution prevention and sustainability. Scientists and engineers have been seeking for ways to make nanotechnology beneficial to the environment. This has been branded as "green nanotechnology" since it promotes technologies that will ensure minimal environmental impact. It may have the potential to address major global sustainability challenges.

Although nanotechnology and nanomaterials permeate all industries, there is no "nanotechnology" industry because it is an enabling technology. The production of nanomaterials is expensive, labor-intensive, and potentially hazardous to the environment. There is a need to develop environmentally friendly methods that are safe for the environment and

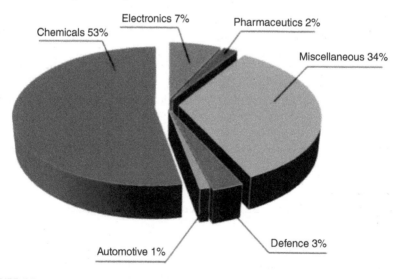

FIGURE 6.1
Some sectors covered by nanotechnology [3].

cost-effective as well. GN aims to produce nanomaterials without deteriorating the environment or human health [4].

6.3 Features of Green Nanotechnology

GN is essentially an application of nanotechnology that envisions sustainability. It is basically making GN products and using them as a support to sustainability without compromising human health. It is the study of how nanotechnology can benefit the environment. There is a strong connection between nanotechnology and the principles and practices of green chemistry and green engineering. GN has two separate but related goals: (1) to produce nanostructures without affecting the environment or human health and (2) to produce nano products that are more environmentally friendly throughout their lifecycle. This is achieved by using eco-friendly materials, recycling after use, and using less energy during manufacture.

GN integrates the principles of green chemistry and green engineering to produce eco-friendly and safe nanomaterials and nano products [5]. Green chemistry, also known as environmentally benign chemistry, is the kind of chemistry that seeks to minimize pollution, conserve energy, and promote environmentally friendly production. It focuses on the design of chemical processes and products to minimize their inherent hazard. The main goal of green chemistry is to reduce or eliminate waste in the manufacture of chemicals and its allied products [6]. Green engineering is an environmentally friendly engineering. It is the design of processes and products that minimize pollution, promote sustainability, and protect human health without sacrificing economic viability and efficiency. It involves creating healthy living environments that use natural resources wisely and conservatively. Green chemistry is an integral part of green engineering since it provides the foundation on which to build green engineering [7]. The key goals of green chemistry and green engineering include fostering sustainability. A marriage of green chemistry and nanotechnology can lead to an emerging new field of "green nano" which holds unprecedented potential toward building an environmentally sustainable society.

6.4 Principles of Green Nanotechnology

The main goals of GN is to produce nanomaterials that do not harm the environment and are eco-friendly. GN uses existing principles of green chemistry and green engineering to make nanomaterials and nano products. Its main principles [8] are as follows:

1. Prevention of generation of wastes; to a great extend it is better to prevent waste than to treat or clean up waste after it is formed.

2. Atom economy, that is, the conversion efficiency of a chemical process in terms of all atoms involved.

3. Less hazardous product synthesis should be there.

4. Safer chemicals and safer solvents should be designed to preserve efficacy of function while reducing toxicity.

5. Reduce derivatives, that is, unnecessary derivatization should be avoided whenever possible.

6.5 Applications

GN includes making and using green nano products in support of sustainability. It has enormous impacts on sciences, engineering, smart electronics materials, medicine, energy, and environmental restoration technologies. Its applications include green manufacturing, water treatment, renewable energy, environmental remediation, agriculture, medical biology, environment, and chemical substitution.

- *Green manufacturing:* GN uses nanotechnology to make manufacturing processes more environmentally friendly and more energy efficient. Using nanotechnology for manufacturing will reduce waste drastically and use less energy during the manufacturing process. A fabrication of organic nanostructures is now possible using GN.

- *Automotive industry:* This is one of the primary exponents of GN. The combination of nanomaterials and nanomanufacturing practices by the automotive industry has turned into a success story for GN. For example, Toyota has been promoting its use of nanocomposite materials for manufacturing energy-efficient vehicles [9].

- *Water treatment:* Producing clean drinking water is a common challenge worldwide. Nanomaterial-based technologies can create eco-friendly solutions for water treatment. Nanotechnology-based water treatment techniques are more economical and efficient than the present, conventional ones.

- *Renewable energy:* There are two major types of energy sources—renewable and nonrenewable sources. Nonrenewable sources release harmful gases to the environment and cause pollution. Renewable are clean or nonpolluting. It is well known that fossil fuel is associated with greenhouse emission and is unsustainable. Renewable energy sources are urgently needed. GN can be used to design solar cells. Solar cell materials that have been suggested include nanoparticles such as titanium dioxide, cadmium telluride, and quantum dots. Renewable energy composed of solar, wind, and biomass will be realized to make the clean energy sources available [10, 11].

• *Modern agriculture:* Today, agriculture faces some challenges such as managing food for the constantly increasing global population. Application of green nanomaterials in agriculture produces eco-friendly products. It reduces costly inputs for plant protection and increase crop productivity [12]. It could be a potential way of enhancing agricultural production. It has influence in every aspect of agriculture. Potential applications of nanotechnology in agriculture are shown in Figure 6.2 [13].

• *Food industry:* Food technology is one of the industry sectors where evolutionary GN will play an important role in the future. Nanofoods (foods produced using nanotechnology) have great potential benefits. Foods and food supplements containing added nanoparticles are becoming available worldwide. Nanotechnology involves using biological molecules for nanostructures that could be used as biosensors on foods. Such biosensors could serve as detectors of food pathogens and as devices to track food products [14].

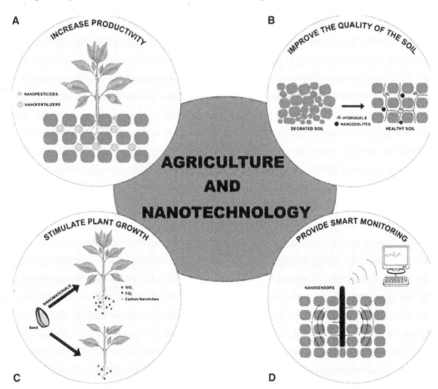

FIGURE 6.2
Potential applications of nanotechnology in agriculture [13].

- *Environmental remediation:* Nanoremediation is the use of nanoparticles for environmental remediation. Nanoremediation has been most widely used for treating wastewater, ground water, soil, sediment, and other environmental contaminants. A corollary principle of GN is waste prevention.

- *Energy:* In today's economy, reliable, efficient, pollution free, abundant energy requirement is the major challenge. Nanomaterials have the potential to change the way we generate, deliver, and use energy. They can be used in making more efficient solar cells and catalysts that can be used in hydrogen-powered fuel cells. Nanotechnology has made big impact on fuel cells, enabling them to convert chemical energy directly into electricity.

- *Medical biology:* Nanoparticles can be applied in several areas of medical biology such as catalysis, cancer treatment, target drug delivery, and biosensors. Because nanoparticles are so small, when inhaled or injected, these tiny structures circulate through the bloodstream and deposit in organs and tissues, where they can build up [15].

Some of these applications are illustrated in Figure 6.3 [16]. These applications will impact a large range of economic sectors, such as energy production and storage and clean-up technologies. Other applications of GN include

FIGURE 6.3
Some applications of green nanotechnology [16].

greener cars, nanofibrillar cellulose, semiconductor/electronics industries, communications, cosmetics, chemicals, materials, green building materials, coatings, tissue engineering, industrial processes, optical science, and carbon nanotubes [17].

6.6 Benefits and Challenges

GN promises to improve our energy economy and provides technological solutions to our global ecological problems. It is generally believed that nanotechnologies will have a significant impact on developing "green" technologies with considerable environmental benefits. It is environmentally friendly and innovative. Nanotechnology enables us to realize material with a wide variety of properties that were previously impossible, impractical, or too expensive. This makes them attractive for scientists and engineers.

GN should not only provide green solutions, but should also "become green" by paying attention to occupational safety and health. Exploring at nano level and understanding the basic building blocks of the materials (tiny structures) that make up our world will result in spectacular breakthroughs. Some of the major benefits of GN are illustrated in Figure 6.4 [18].

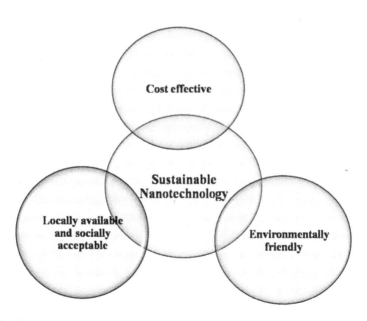

FIGURE 6.4
Some of the benefits of green nanotechnology [18].

Although GN poses many advantages over traditional methods, there is the concern about the potentially negative long-term effects of engineered nanoparticles on human health and the environment. The opportunities offered by GN solutions should be balanced with a number of practical challenges, environmental and social issues, and health and safety concerns [19].

Nanotechnology is becoming more complex and it is both multidisciplinary and interdisciplinary. Another concern is whether the legal authorities and governance tools are adequate to address the potential risks posed by nanotechnology. There is an ethical and social responsibility on the production and use of nanomaterials and their potential environmental effects.

6.7 Conclusion

GN, as an emerging technology, is the study of how nanotechnology can be applied to benefit the environment. It integrates the principles and practices of green chemistry and green engineering to produce eco-friendly, safe, nanostructures that do not use toxic chemicals. It is a rapidly developing field at the interface of chemistry, physics, engineering, and biology. It is one of our key future issues.

Although GN is still in its infancy, its rapid rise has led some technologists to call it the next industrial revolution. Collaboration of academic researches and private sector can lead to better understanding of the potential risks of nanotechnologies and how to address them. This will also help in successfully moving green nanotechnologies to the manufacturing facility.

Education plays a major role in the successful diffusion of GN. Unfortunately, very few universities are involved in nanotechnology. The government should do more to promote the wide adoption of green nanotech and to make United States a global leader in GN. Small firms as well as international organization play a crucial role in the diffusion of the new nanotechnologies [20]. More information about GN can be found in the books in References [3, 21–30] and also in the journals devoted to nanotechnology:

- *Nanotechnology*
- *International Journal of Green Nanotechnology*
- *Journal of Nanoscience and Nanotechnology*
- *Journal of Nanotechnology and Materials Science*
- *International Journal of Nano Dimension*
- *International Journal of Green Nanotechnology: Physics and Chemistry*

References

1. M. N. O. Sadiku, M. Tembely, and S. M. Musa, "Nanotechnology: An introduction," *International Journal of Software and Hardware Research in Engineering*, vol. 4, no. 5, May, 2016, pp. 40–44.

2. E. D. Sherly, K. Madgular, and R. Kakkar, "Green nanotechnology for cleaner environment present and future research needs," *Current World Environment*, vol. 6, no. 1, 2011, pp. 177–181.

3. O. Figovsky and D. Beilin, *Green Nanotechnology*. Singapore: Pan Stanford Publishing, 2017, p. xv.

4. S. Srivastava et al., "Green nanotechnology," *Journal of Nanotechnology and Materials Science*, vol. 3, no. 1, 2016, pp. 17–23.

5. "Green nanotechnology," *Wikipedia*, the free encyclopedia https://en.wikipedia.org/wiki/Green_nanotechnology

6. M. N. O. Sadiku, S. M. Musa, and O. M. Musa, "Green chemistry: A primer," *Invention Journal of Research Technology in Engineering and Management*, vol. 2, no. 9, September 2018, pp. 60–63.

7. M. N. O. Sadiku, S. R. Nelatury, and S. M. Musa, "Green engineering: A primer," *Journal of Scientific and Engineering Research*, vol. 5, no.7, 2018, pp. 20–23.

8. A. Goel and S. Bhatnagar, "Green nanotechnology," *BioEvolution*, January 2014, pp. 3–4.

9. C. Milburn, "Greener on the other side: Science fiction and the problem of green nanotechnology," *Configurations*, vol. 20, no. 1, January 2012, pp. 53–87.

10. K. W. Guo, "Green nanotechnology of trends in future energy: A review," *International Journal of Energy Research*, vol. 36, 2012, pp. 1–17.

11. I. Khan and A. Sharma, "Nanotechnology for sustainable development: Reducing carbon emissions through clean energy technologies," *International Journal of Systems, Algorithms & Applications*, vol. 2, no. 5, May 2012, pp. 29–31.

12. P. Ashoka, "Book review: Green nanotechnology is a key for eco-friendly agriculture," *Journal of Cleaner Production*, vol. 142, 2017, pp. 4440–4441.

13. L. F. Fraceto et al., "Nanotechnology in Agriculture: Which Innovation Potential Does It Have?" https://www.frontiersin.org/articles/10.3389/fenvs.2016.00020/full

14. R. Ravichandran, "Nanotechnology applications in food and food processing: Innovative green approaches, opportunities and uncertainties for global market," *International Journal of Green Nanotechnology: Physics and Chemistry*, vol. 1, no. 2, 2010.

15. D. Nath and P. Banerjee, "Green nanotechnology—A new hope for medical biology," *Environmental Toxicology and Pharmacology*, vol. 36, 2013, pp. 997–1014.

16. S. P. Goutam et al., "Green synthesis of nanoparticles and their applications in water and wastewater treatment," https://www.researchgate.net/publication/328051538_Green_Synthesis_of_Nanoparticles_and_Their_Applications_in_Water_and_Wastewater_Treatment/download

17. A. Goel and S. Bhatnagar, "Green nanotechnology," *BioEvolution*, January 2014, pp. 3–4.

18. S. Saif, A. Tahir, and Y. Chen, "Green synthesis of Iron nanoparticles and their environmental applications and implications," *Nanomaterials*, vol. 6, no. 11, 2016.

19. I. Iavicoli et al., "Opportunities and challenges of nanotechnology in the green economy," *Environmental Health*, vol. 13, 2014.

20. A. Glaser et al., "Enabling nanotechnology entrepreneurship in a French context towards a holistic theoretical framework," *Journal of Small Business and Enterprise Development*, vol. 23, no. 4, 2016, pp. 1009–1031.
21. G. B. Smith and C. G. Granqvist, *Green Nanotechnology; Solutions for Sustainability and Energy in the Built Environment*. Boca Raton, FL: CRC Press, 2011.
22. M. L. Larramendy and S. Soloneski (eds.), *Green Nanotechnology—Overview and Further Prospects*. Rijeka, Croatia: InTech, 2016.
23. M. Nasrollahzadeh et al., *An Introduction to Green Nanotechnology*. Cambridge, MA: Academic Press, 2019.
24. N. Srivastava et al. (eds.), *Green Nanotechnology for Biofuel Production*. Berlin: Springer, 2018.
25. M. Nasrollahzadeh et al. (eds.), *An Introduction to Green Nanotechnology*. Cambridge, MA: Academic Press, 2019.
26. G. Blokdyk, *Green nanotechnology: Practical Tools for Self-Assessment*, 3rd ed. 5STARCooks, 2018.
27. V. A. Basiuk and E. V. Basiuk (eds.), *Green Processes for Nanotechnology: From Inorganic to Bioinspired Nanomaterials*. Berlin: Springer, 2015.
28. N. Dasgupta, S. Ranjan, and E. Lichtfouse, *Environmental Nanotechnology*. Berlin: Springer, 2019.
29. O. Figovsky and D. Beilin, *Green Nanotechnology*. Boca Raton, FL: CRC Press, 2017.
30. M. H. Fulekar and B. Pathak, *Environmental Nanotechnology*. Boca Raton, FL: CRC Press, 2017.

7

Green IT/Computing

The person who won't read has no advantage over one who can't read.

—**Mark Twain**

7.1 Introduction

In the 21st century, the word "green" has evolved to relate to environmental issues. Behaving in an environmentally sound way will be essential to our future. The earth may no longer be a sustainable living environment for any creature if we do not reduce the rate and amount of toxic waste [1]. High level of carbon dioxide (CO_2) emission is dangerous and can cause health problems.

Computing technology has become an essential part of global infrastructure due to worldwide usage of computer devices. In this modern era, all sectors of the economy like business, medical, education, transportation, and agriculture use computers and electronic devices, which somehow cause harm to the environment [2]. As computing becomes increasingly pervasive, the energy consumption due to computing keeps increasing.

Green computing (GC), also called green technology or green IT, is emerging as a critical information communication technology to reverse the trend. The terms "green computing," "green IT," and "sustainable computing" are often used interchangeably. GC refers to the practice of reducing environmental footprints of technology by efficiently using computing assets. It comprises strategies and best practices for optimizing the usage of computing resources and reducing the environmental footprint of technology. It is a critical building block for corporate social responsibility. It uses computer systems in a sustainable environmental way that minimizes power usage of electronic devices/systems like monitors, desktops, printers, communication gadgets, and data centers [3]. As concern for climate change and sustainability continues to grow, businesses around the world are realizing that green IT initiatives offer cost-saving benefits.

This chapter provides a brief introduction to green IT/computing. It first addresses the general idea of greening technology and then brings out the concept of GC. It explains ways to implement GC and its benefits and challenges. Then it discusses current trends in green IT/computing. The last section concludes this chapter.

7.2 Pathways to Greening

One of the biggest challenges facing the environment today is global warming due to carbon emissions. The term "green" is used to refer to environmentally sustainable activities. It means using computers in ways that save the environment, energy, and money.

The concept of green IT/computing emerged in 1992 when the U.S. Environmental Protection Agency launched Energy Star. Energy Star served as a voluntary award given to manufacturers who succeeded in minimizing energy consumption while maximizing efficiency. Energy Star program is still active. It reduces the amount of energy consumed by a product by automatically switching it into "sleep" mode when not in use. This is a labeling standard for energy efficiency in electronic equipment. Energy Star was applied to different types of devices such as computer monitors, television sets, refrigerators, air conditioners, etc. [4]. Since 1992 many governmental agencies and nonprofit organizations have implemented standards and regulations that encourage GC. Some countries have launched a number of "paperless" initiatives with the aim of reducing the use of paper in offices [5]. Many companies are exploring methods and developing policies to use green technologies. Companies in the computer industry are interested in GC because it saves energy and expenditure cost. GC is illustrated in Figure 7.1 [6].

Green IT can also be viewed from an Information Systems (IS) perspective termed "green IS." Green IS involves people and their use of IT. While green IS is the use of information systems to achieve environmental objectives, green IT emphasizes reducing the environmental impacts of IT production and use. With the recent developments in IT, business organizations are being made aware of their environmental effects and social responsibility [7].

To be green requires [8]: (1) improving energy efficiency by reducing carbon footprint, (2) reducing e-waste, and (3) enabling lifestyle changes that lower impact on the environment. The goals of GC are similar to those of green chemistry: reduce the use of hazardous materials and maximize energy efficiency during the product's lifetime. The objective of GC is to save the environment from the harmful effects of computers, servers, monitor, printers, and communications systems. It involves saving energy or reduction of carbon footprints. GC is the movement toward a more environmentally sustainable computing. It seeks to conserve the energy and reduce the e-waste. It is important for all classes of computing systems, from handheld mobile devices to data center facilities, which are heavy consumers of energy [9, 10].

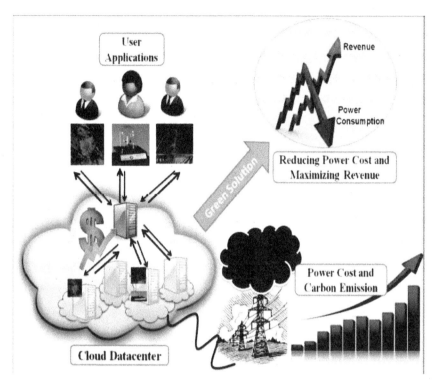

FIGURE 7.1
A typical illustration of green computing [6].

7.3 Tips for GC Users

Although we realize that GC is important, some people just do not do what it takes to save our earth. The work habits of computer users and businesses can be modified to minimize adverse impact on the global environment. Here are some tips on what you can do to make computing usage more green [11].

- Turn off computers, printers, etc., at the end of each day or when not in use.
- Use sleep mode when away from a computer.
- Forgo the screen saver—it does not save energy.
- Consider using an inkjet printer.
- Refill printer cartridges, rather than buying new ones.
- Refurbish an existing computer instead of buying a new one.
- Use liquid-crystal-display (LCD) monitors rather than cathode-ray-tube (CRT) monitors.

- Use email communications as an alternative to paper memos and fax documents.
- Minimize consumption of paper and properly recycle waste paper.
- Minimize the use of electricity or energy.
- Use more renewable energy sources.
- Use devices that consume less energy.
- Use less hazardous materials in computing devices.
- Ensure proper disposal of electronic waste.
- Buy Energy star–labeled products.

We must be responsible and do our part to reduce the environmental impact of computing.

7.4 Green IT/Computing Strategies

GC is also known as green information technology (green IT). It may be regarded as the practice of using computing resources to achieve maximum productivity with no harmful effects on the environment. It comprises the measures and strategies designed to reduce our environmental impact. It is the use of IT resources in an energy-efficient and cost-effective manner. It is green principles and practices applied to IT to achieve environmental sustainability. It relates to the ability of an organization to reduce CO_2 emissions, decrease energy consumption, and minimize e-waste. IT may be regarded as both a problem and a solution for environmental sustainability, or both a barrier and an enabler of environmental sustainability. Green IT involves both using IT to reduce our environmental impact and reducing the environmental impact of the IT sector. Numerous IT applications, such as e-commerce, smart grids, smart buildings, digital media, and intelligent transport systems, have a positive effect on reducing environmental pollution and carbon emissions [12].

The principal goal of green IT/computing is to provide awareness to the public to help sustain an eco-friendly environment. The key objectives of GC include [13]:

- Minimizing energy expenditure
- Purchasing green energy
- Reducing paper and other consumables used
- Minimizing equipment removal necessities
- Reducing travel requirements for employees/customers

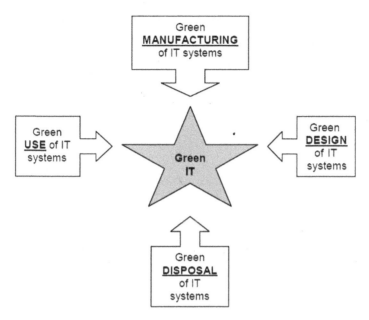

FIGURE 7.2
Four key areas in green IT/computing [14].

The green IT/computing principles show the concepts of reducing the environmental impact. They consist of four key green holistic principles, which are illustrated in Figure 7.2 and explained as follows [14].

- *Green use:* Reduce the energy consumption of data centers and computer-related devices. IT practitioners should deploy green usage initiatives to reduce energy utilization of computers and other IT infrastructure.

- *Green design:* Design energy efficient and environmentally sound components, computers, servers, and cooling equipment. Green design aims to decrease use of nonrenewable resources, manage nonrenewable resources, and reduce toxic emissions.

- *Green manufacturing:* Manufacture electronic components, computer-related subsystems with minimal or no impact on the environment. Every process in manufacturing should have a low or no impact on the environment.

- *Green disposal:* Green disposal aims to reduce e-waste by repairing, redeploying, or disposing, refurbishing, retaining, and reusing of outdated IT hardware. In this practice, the company should plan to refurbish and reuse unwanted computers or other electronics components.

From this, we notice that the main objective of GC involves the usage, designing, and disposing computer devices, and manufacturing of information and communication technology (ICT) in a way that reduces their environmental impact. Therefore, a green computer is one in which the entire process from design, manufacture, use, and disposal involves as little environmental impact as possible.

7.5 Implementing Green IT/Computing

Green IT/computing is currently becoming an important topic for organizations. Although green IT is gaining popularity, not every organization is ready to implement it. Therefore, it is necessary to consider ways to implement green IT practices in any organization. There is no easy way to green IT/computing. We must strive to minimize greenhouse gases and waste, while increasing the effectiveness of IT, such as computers, data centers, and computer networks. A GC activity or initiative must cover all territories: people, organizations, equipment, and networks.

1. *Home usage:* The home office is an area of the house where energy is wasted and lost. Buy Energy Star–labeled monitors, desktops, laptops, and printers. Turn off your PC, laptops, and other equipment or put them in "sleep" mode when not in use. Automatic switching off of the computers when they are not in use helps reduce the power consumption. E-cycle used electronic equipment. Minimize using papers through the paperless approach and reduce the use of printers.

2. *E-waste recycling:* This is recycling of e-waste such as old computers, monitors, phone, and TV. One can donate old computer-related products to schools and nonprofit organizations instead of throwing them away. Much electronic waste can be recycled, the parts used to make new items. Recycling computing hardware can keep unsafe materials (such as lead and mercury) out of landfills. Proper management of e-waste is a good potential route to implementing GC [15].

3. *Energy-efficient data centers:* Data center facilities are harmful for the environment and are known for their energy hunger, energy inefficiency, and wasteful energy consumption. Large amounts of energy were needed for powering servers and cooling them, thereby incurring enormous carbon footprints. Energy consumption of data centers can be minimized by applying various power management techniques to both switching and networking components. The

U.S. Department of Energy specifies five primary areas on which to focus energy-efficient data center design best practices: IT systems, environmental conditions, air management, cooling systems, and electrical systems. It will require green cloud computing solutions to save energy and reduce operational costs and carbon emission [16].

4. *Organizations:* Organizations are causing an increase in the carbon footprint. They should have a budget for GC. They should have policies for printing, recycling, and IT procurement. An international consortium of computer companies, including IBM, Dell, and Sun Microsystems, organized The Green Grid in 2007 to improve energy efficiency in business computing systems [17]. To encourage paperless policy, many universities charge for printing, which makes students take some responsibility for their printing. Create awareness for green IT in the day-to-day activities and operation of the organization. Have green IT policy that includes definition and vision of green IT, strategic plans for green IT, etc. Inform staff and practitioners on the ethics of sustainability.

5. *Virtualization:* One can reduce the number of servers and the corresponding energy consumption by using virtualization technology. Virtualization refers to the abstraction of computer resources. It allows a system administrator to combine two or more computer systems on one set of physical hardware. It is a technique of creating virtual version of real things like the hardware, software, memory, data, network, etc. By using virtual software to perform tasks, a single server can be used to power these virtual servers and reduce energy consumption [18]. Today, virtualization technology has become an essential tool to provide resource flexibly for each user. Besides virtualization, another energy-saving technique is cloud computing.

6. *New technologies:* Technology has a complex relationship with environmental issues. By going green in technology, we help promote an eco-friendly and cleaner environment. New technologies such as cloud computing and fog computing are a solution for reducing energy consumption. Cloud computing addresses two main challenges related to GC: energy usage and resource consumption. A new supercomputer, made in Germany, emerged as the most energy-efficient supercomputer in the world. GC advocates the efficient usage of computing resources in all emerging technologies, such as big data, cloud computing, mobile computing, and Internet of things.

Green IT/computing can be implemented as a software, hardware, or business process. Implementing green IT reduces the organization's impact on the environment.

7.6 Benefits and Challenges

The main benefits of green IT can be classified into three areas [19]: energy efficiency of IT, eco-compatible management of the lifecycle of IT, and IT as an enabler of green governance. By adopting green IT solutions, companies save a lot of money and help minimize their environmental impact. Reducing hardware implies less energy usage because there are fewer devices. Today the most talented people in IT prefer to work with organizations that are considered environmentally progressive.

GC can enable computer systems, people, society, and environment in better harmony. By choosing GC hardware and software resources, we can significantly reduce carbon footprint, save money, and improve the reuse cycle. Other benefits of GC include [20]:

- Reduction in energy consumption of computing resources.
- Reduction in carbon emission.
- Reduction in harmful effects of computing resources.
- Reduction in computing wastes
- Reduction in operational cost.
- Saving of energy during idle operation.
- Use of eco-friendly sources of energy.
- Use of resources such as computers, data centers, light, etc. in an environmental-friendly way.
- Improvement in corporate image by meeting compliance and regulatory requirements.
- Strong returns of investment (ROI).

GC presents some challenges for business people, engineers, and architects.

GC requires that designer take the product life cycle into consideration, from production to operation to recycling. The capital cost of implementing green IT (replacing existing hardware and software with greener IT products) is often regarded as the main challenge. Reengineering of business processes and practices may be a major stumbling block. Redesigning an entire company to leverage green IT might be risky since it challenges some of the ongoing day-to-day operations. Reluctance to change (cultural and behavioral) is another key barrier. New emerging applications such as smart homes, transportation, healthcare, manufacturing, home automation, and power grid bring new challenges to meet sustainability requirements. There are privacy and ethical issues that arise from the recycling of the old computer. Computers gathered through recycling drives are often shipped to developing countries, where environmental standards are less strict than in Western world. Developed countries are already implementing GC solutions, while developing countries are just at awareness stage. There is a need for green metric set that will allow for measuring and monitoring energy consumed and wasted in computing. Other barriers include inadequate funding, misalignment with physical facilities, and a lack of resources such as IT staff.

7.7 Current Trends on Green IT/Computing

Current trends of GC are toward efficient utilization of resources. There are various ways by which researchers are seeking to achieve this [21]. The most popular approaches are using virtualization and cloud computing.

A major trend of GC is virtualization of computer resources. Virtualization may be the easiest way to implement green IT in any organization because it provides a better utilization of computer hardware. Virtualization allows the design of several virtual machines on one physical machine, thereby increasing capacity utilization of physical servers and reducing energy need. Virtualization helps make large strides in GC technology. It provides a path toward green by offering virtualization software as well as management software. Virtualization and green solutions both are easier to implement if one has a well-understood hardware base.

Cloud computing is the latest trend in the field of GC. It does away with the hardware servers and uses virtual servers. Thus, cloud computing is energy-efficient technology for ICT. It provides better resource utilization, which is good for the sustainability movement for green technology [22].

The use of virtualization along with that of cloud computing is playing an important role in GC concept. Cloud virtualization involves virtualizing not just resources but also the location and ownership of the infrastructure through the use of cloud computing. It leverages the economies of scale due to large numbers of organizations sharing the same infrastructure.

7.8 Conclusion

GC is the practice of using computing resources efficiently. It is ecologically sustainable computing with the goal of increasing the energy efficiency during the product lifetime. By going green in computing, we help advance an eco-friendly and cleaner environment. As computing becomes increasingly pervasive, GC has been a major concern for businesses and governments worldwide.

Today, organizations have come to realize that going green is in their best interest and are moving toward implementation of GC platform. Several

IT departments have GC initiatives to reduce the environmental effect of their IT operations. Many governmental agencies have started to implement standards and regulations to support GC. Many universities are also launching academic initiatives in this discipline [23]. In order for the whole society to consider GC an essential part of environmental responsibility, the average citizen (including college students) must be well informed. Awareness should also be increased regarding GC since it will have major impact in future computing.

By eliminating the environment-unfriendly aspects of computer systems, GC can enable computer systems, computer networks, people, society, and environment to be in better harmony. The adoption of GC requires that IT users be well informed about the various facets of the notion. Although green IT/computing is a relatively young field, its impact has already changed lots of commonly accepted human activities. Companies are willing to spend more today to implement green technology than in the past. Companies that implement green IT solutions save a lot of money and reduce their environmental impact. More information on green IT/computing can be found in the books in [24–28] and the journal exclusively devoted to it: *International Journal of Green Computing*.

References

1. W. S. Chow and Y. Chen, "Intended belief and actual behavior in green computing in Hong King," *The Journal of Computer Information Systems*, vol. 50, no. 2, Winter 2009, pp. 136–141.
2. A. Chopra, S. Sharma, and V. Kadyan, "Need of green computing to improve environmental condition in current era," *Proceedings of the International Conference on Electrical, Electronics, and Optimization Techniques*, 2016, pp. 3209–3217.
3. M. N. O. Sadiku, N. K. Nana, and S. M. Musa, "Green computing: A primer," *Journal of Scientific and Engineering Research*, vol. 5, no. 4, 2018, pp. 247–251.
4. C. Joumaa and S. Kadry, "Green IT: Case studies," *Energy Procedia*, vol. 16, 2012, pp. 1052–1058.
5. M. A. Aljaberi, S. N. Khan, and S. Muammar, "Green computing implementation factors: UAE case study," *Proceedings of the 5th International Conference on Electronic Devices, Systems and Applications*, 2016.
6. S. Vikram, "Green computing," *Proceedings of the International Conference on Green Computing and Internet of Things*, October 2015, pp. 767–777.
7. M. N. O. Sadiku, P. O. Adebo, and S. M. Musa, "Green IT: A primer," *International Journal of Advanced Research in Computer Science and Software Engineering*, vol. 8, no. 10, October 2018, pp. 4–6.
8. S. Singh, "Green computing strategies & challenges," *Proceedings of the International Conference on Green Computing and Internet of Things*, 2015, pp. 758–760.

9. "Green computing," *Wikipedia*, the free encyclopedia https://en.wikipedia.org/wiki/Green_computing

10. B. Saha, "Green computing," *International Journal of Computer Trends and Technology*, vol. 14, no. 2, August 2014, pp. 46–50.

11. G. Appasami and J. K. Suresh, "Optimization of operating systems towards green computing," *International Journal of Combinatorial Optimization Problems and Informatics*, vol. 2, no. 3, September–December 2011, pp. 39–51.

12. Q. Deng and S. Ji, "Organizational green IT adoption: Concept and evidence," *Sustainability*, vol. 7, 2015, 16737–16755.

13. D. R. R. Joan, "Introduction to green computing model for clouds," *i-manager's Journal on Cloud Computing*, vol. 1, no. 3, May–July 2014, pp. 1–7.

14. S. Murugesan, "Harnessing green IT: Principles and practices," *IEEE IT Professional*, January–February 2008, pp. 24–33.

15. B. Debnath, R. Roychoudhuri, and S. K. Ghosh, "E-waste management – A potential route to green computing," *Procedia Environmental Sciences*, vol. 35, 2016, pp. 669–675.

16. S. Rawat et al., "An analytical evaluation of challenges in green cloud computing," *Proceedings of the International Conference on Infocom Technologies and Unmanned Systems (Trends and Future Directions)*, December 2017, pp. 351–355.

17. P. Kurp, "Green computing," *Communication of the ACM*, vol. 51, no. 10, October 2008, pp. 11–13.

18. Anju, "Security issues in green computing," *International Journal of Advanced Research in Computer Science*, vol. 8, no. 4, May 2017, pp. 157–160.

19. M. Thomas, D. Costa, and T. Oliveira, "Assessing the role of IT-enabled process virtualization on green IT adoption," *Information Systems Frontiers*, vol. 18, no. 4, August 2016, pp. 693–710.

20. G. Jindal and M. Gupta, "Green computing 'future of computers," *International Journal of Emerging Research in Management &Technology*, December 2012, pp. 14–18.

21. A. Shibly, "Green computing: Emerging issue in IT," https://www.researchgate.net/publication/276266707_Green_Computing_-_Emerging_Issue_in_IT

22. R. Pandey et al., "The rising era of green computing," *International Journal of Computer Science and Mobile Computing*, vol. 6, no. 2, February 2017, pp. 127–130.

23. P. K. Paul et al., "Green and environmental friendly domain and discipline: Emerging trends and future possibilities," *International Journal of Applied Sciences & Engineering*, vol. 2, no. 1, April 2014, pp. 55–67.

24. B. E. Smith, *Green Computing: Tools and Techniques for Saving Energy, Money, and Resources*. Boca Raton, FL: CRC Press, 2014.

25. G. Hart-Davis, *The Healthy PC: Preventive Care, Home Remedies, and Green Computing*, 2nd ed. McGraw-Hill, 2017.

26. P. O. de Pablos (ed.), *Green Technologies and Business Practices: An IT Approach*. Information Science Reference, 2013.

27. T. Velte, A. Velte, and R. C. Elsenpeter, *Green IT: Reduce Your Information System's Environmental Impact While Adding to the Bottom Line*. New York, NY: McGraw-Hill, 2009.

28. W. C. Feng (ed.), *The Green Computing Book: Tackling Energy Efficiency at Large Scale*. Boca Raton, FL: CRC Press, 2014.

8

Green Internet of Things

Men of genius are admired, men of wealth are envied, men of power are feared, but only men of character are trusted.

—Zig Ziglar

8.1 Introduction

The Internet technology has provided many beneficial applications for our daily lives. It has played a major role in flourishing every sector of the economy. Recently, we have witnessed more and more devices interconnected through the Internet, creating Internet of Things (IoT). The IoT allows people and things to be connected anywhere, anytime, with anyone, and anything. It bridges the gap between the cyber world and the physical world, enabling billions of connected devices to communicate. It may be regarded as the advance version of machine-to-machine (M2M) communication, where every object connects with another object without human intervention.

The aim of IoT is to connect devices or things (e.g., cars, sensors devices, mobile phones, cloud computing systems, people, cameras, social networks, the radio-frequency identification (RFID) network, the Global Positioning System (GPS) network, the 5G network) from the physical world to the cyber world and let them interact with each other. These connected heterogeneous devices tend to generate a massive volume of big data. They also consume a substantial amount of energy. The IoT not only consumes energy, it causes toxic pollution and e-waste. Thus, the greenness of IoT is critically important for the success of IoT. Green IoT focuses on reducing the energy consumption of IoT, thereby fulfilling the smart world with sustainability. Both the devices and the protocols used in communicating should be energy efficient. Thus, green IoT is regarded as the future of IoT that is environmentally friendly.

This chapter provides an introduction to green IoT. It begins by providing a brief overview of IoT and the technologies enabling green IoT. It presents some applications and addresses some challenges facing green IoT. The last section concludes the chapter.

8.2 Overview of Internet of Things

IoT is a newly emerging concept, which aims to connect billions of devices with each other. The IoT comprises a number of technologies that enable global connectivity over the worldwide physical objects. IoT is the next step in the evolution of the Internet since it takes into consideration all devices connected to it. It allows all types of elements (sensors, actuators, personal electronic devices, laptops, tablets, digital cameras, smart phones, alarm systems, home appliances, industrial machines, etc.) to autonomously interact with each other. It allows things to be connected, sensed, and collaboratively communicate over the Internet. Integration of every device with the Internet necessitates that devices use an Internet Protocol (IP) address as a unique identifier. To some extent, the future of IoT will be limited without the support of IPv6.

There are four main technologies that enable IoT [1]: (1) RFID and near-field communication, (2) optical tags and quick response codes, (3) Bluetooth low energy (BLE), and (4) the wireless sensor network. Others include biometrics and nanotechnologies. These technologies enable connected items in IoT to sense the physical environment, collect or transfer data, and communicate with other things.

Applications of IoT include smart grid, smart cities, smart manufacturing, education, e-health, food supply chains, and intelligent transportation. Although the benefits of IoT are great, IoT consumes energy and causes toxic pollution and e-waste. The rechargeable batteries for IoT devices have limited life span and this makes the IoT technology to be on the verge of inadequate battery power. The greenness of IoT is crucial for the success of IoT from the perspective of energy efficiency. Green IoT is regarded as the future of IoT that is environmentally friendly [2].

8.3 Enabling Green IoT

The exchange of a large amount of information among billions of devices connected to IoT creates a massive energy need. Green IoT promises to achieve a lower power consumption than IoT and make the environment safer. The life cycle of green IoT contains green design, green manufacturing, green utilization, and green disposal with a minimal impact on the environment, as shown in Figure 8.1 [3]. Monitoring the waste is crucial for proper recycling.

The five technologies enabling green IoT are [4]: green RFID, green wireless sensor network (WSN), green cloud computing (CC), green data center (DC), and green M2M.

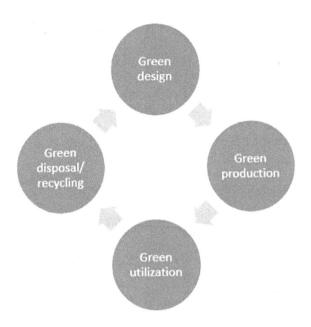

FIGURE 8.1
Life cycle of green IoT [3].

Green RFID: The RFID is a small electronic device that includes several RFID tags and a small tag reader. The tags are used for storing information regarding the objects to which they are attached. Green RFID may mean reducing the size of RFID tags.

Green WSN: WSN consists of sensor nodes that are equipped with sensors that take readings (e.g., temperature, humidity, CO_2 detector, soil control, etc.) from the surroundings. Green WSN may imply making sensor nodes only work when necessary and using data and context-awareness algorithms to reduce the data size.

Green CC: CC offers three kinds of services for different applications: IaaS (Infrastructure as a Service), PaaS (Platform as a Service), and SaaS (Software as a Service) and delivers computing as a utility. With applications being moved to the cloud, more power is consumed causing CO_2 emissions. The green CC concept has an important role in reducing energy consumption in IT industries. Green CC is becoming important due to the increasing concerns about environmental issues by cloud service providers. Data management and efficient infrastructure are critical to facilitate green CC.

Green DC: DC consume a lot of energy with high operational costs and large CO_2 footprints. Improving energy efficiency of a DC may require using renewable energy or green sources of energy (such as wind, water, solar energy, heat pumps).

Green M2M: M2M is a technology that allows both wireless and wired devices to communicate with other devices of the same type. It allows machines to consume the information other machines generate. It involves massive machines and consumes a lot of energy. Green M2M may involve switching some nodes to low-power operation/sleeping mode.

Also, various 5G-enabling technologies play a key role in the reduction of energy consumption of IoT systems. With the advances of these enabling technologies, green IoT has a great potential to enhance economic and environmental sustainability. To achieve green IoT, the following principles should be considered [3]:

- Turn off devices that are not needed.
- Use renewable energy for charging and utilization purposes.
- Use energy-efficient optimization techniques.
- Use data and context-awareness algorithms to reduce the data size.
- Use energy-efficient routing techniques to reduce the mobility power consumption.
- Send only data that are needed.
- Minimize the length of the wireless data path.

To enable a green IoT, the IoT should be characterized by energy efficiency. The energy-efficient procedures (hardware or software) can facilitate reducing the greenhouse effect of existing applications and services [5].

8.4 Applications

IoT and green IoT have a wide range of applications and services. These include smart grid, smart city, smart home, smart healthcare, remote monitoring, agriculture, intelligent transportation, green industrial IoT, and industrial automation [6]. Some of these are illustrated in Figure 8.2 [7].

- *Smart cities:* Smart cities include smart buildings, smart transportation, smart parking system, and smart waste. Green initiatives in IoTs are important, as the smart cities are being developed all across the globe. In the smart city initiative, massive deployment of IoT sensors and devices will enable the implementation of IoT-related services. Energy-efficient IoT systems play a key role in smart cities and smart buildings. Taking green initiatives will make the IoTs efficient in resource utilization. All the components (RFID, sensor network, data center, etc.) need to be green for the overall greenness of the network [8].
- *Smart home:* Green IoT enables home equipped with lighting, heating, and electronic devices to be controlled remotely by a smartphone or a computer. A central computer accepts voice commands,

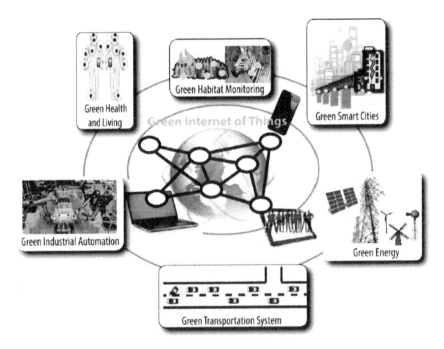

FIGURE 8.2
Green IoT applications [7].

distinguishes between occupants for personalized responses. Personal lifestyle at home is enhanced by making it more convenient and easier to monitor and operate home appliances and systems. For example, based on the weather forecast information, a smart home can automatically lower or close the window blinds [9].

- *Industrial automation:* Energy efficiency has a priority in many corporate agendas across industries. The emergence of IoT in the industrial sector has led to various applications that have helped to reduce the CO_2 emissions and have provided a path for green industrial automation. Smart environments help to improve the automation of industrial plants by using RFID. The RFID reader directly communicates with the machine/robot without any human intervention [6].

- *Climate-smart agriculture:* Modern tools and technologies are needed to improve the production and quality of crops. By promoting new methods and technologies, climate-smart agriculture (CSA) helps farmers to manage their resources, boost their profits, and reduce agriculture's contribution to climate change. Even small-scale farmers in developing nations can achieve success and increase farm production by adopting CSA technologies. CSA will enable farmers to contend with the enormous challenges that they have to deal with [10].

- *Smart healthcare:* The healthcare sector has benefited substantially from IoT technologies. Healthcare applications can be improved by embedding biometric actuators and sensors in patients for capturing, monitoring, and tracking a human body. Smart healthcare may result in efficient healthcare services, enhancing the care quality, improving access to care, and decreasing care costs [2].

The green IoT is predicted to introduce significant changes in our future daily lives and lead to a green environment.

8.5 Benefits and Challenges

Green IoT will contribute toward reducing pollutions, exploiting environmental conservation, minimizing operational costs, optimizing power consumption, and maximizing bandwidth utilization. It poses a great potential to bolster economic and environmental sustainability. It will control critical infrastructure such as smart cities, smart transportation, and smart power grids.

Green IoT faces a lot of challenges that need to be addressed before its full realization. Due to the complex and heterogeneous nature of IoT, various things connected to it consume a high volume of memory, energy, and high bandwidth. Energy efficiency in IoT systems has been a core challenging issue. The energy consumption in IoT can be minimized by wisely adjusting transmission power, activity scheduling, designing energy efficient data centers, energy efficient transmission of data from sensors, and implementing energy efficient policies [11]. Developing green deployment schemes for IoT is challenging due to its complex and heterogeneous nature. Achieving green IoT through the use of 5G poses new challenges due to the need for transferring huge volume of data in an efficient way [12]. Scalability and cybersecurity remain outstanding challenges of green IoT applications. Reliability is a major challenge for achieving green IoT, because not all sensor nodes are expected to be simultaneously active in the IoT domain. Reliability is critical for efficient IoT communication, because unreliable sensing, processing, and transmission can cause false monitoring data reports, which would reduce people's interest in IoT [13].

8.6 Conclusion

The IoT will interconnect every aspect of people's world in the smart world. It will control critical infrastructure such as the smart power grid, smart cities, smart manufacturing, and smart transportation systems. As IoT becomes a pervasive technology, its sustainability and environmental effects are critically important. Green IoT is the IoT technology tasked with enabling a greener society. It promises to improve energy efficiency in IoT systems and help reduce environmental pollution. It is regarded as the future of IoT that is environmentally friendly. Green IoT is still in its infancy and its adoption faces many technical challenges.

References

1. M. N. O. Sadiku, S. M. Musa, and S. R. Nelatury, "Internet of things: An introduction," *International Journal of Engineering Research and Advanced Technology*, vol. 2, no. 3, March 2016, pp. 39–43.
2. S. H. Alsamhi et al., "Greening Internet of things for smart everything with a green environment life: A survey and future prospects," https://arxiv.org/ftp/arxiv/papers/1805/1805.00848.pdf
3. M. A. M. Albreem et al., "Green Internet of things (IoT): An overview," *Proceedings of the 4th IEEE International Conference on Smart Instrumentation, Measurement and Applications*, Putrajaya, Malaysia, November 2017.
4. C. Zhu et al., "Green Internet of things for smart world," *IEEE Access*, vol. 3, 2015, pp. 2151–2162.
5. C. S. Nandyala and H. K. Kim, "Green IoT agriculture and healthcare application (GAHA)," *International Journal of Smart Home*, vol. 10, no. 4, 2016, pp. 289–300.
6. A. Gapchup et al., "Emerging trends of green IoT for smart world," *International Journal of Innovative Research in Computer and Communication Engineering*, vol. 5, no. 2, February 2017, pp. 2139–2148.
7. F. K. Shaikh, S. Zeadally, and E. Exposito, "Enabling technologies for green Internet of things," *IEEE Systems Journal*, vol. 11, no. 2, June 2017, pp. 983–998.
8. S. K. Routray and K. P. Sharmila, "Green initiatives in IoT," *Proceedings of 3rd International Conference on Advances in Electrical, Electronics, Information, Communication and Bio-Informatics*, 2017.
9. J. S. Kumar, "Green smart world (Internet of things)," *International Journal of Engineering Science Invention*, 2018, pp. 32–35.
10. M. N. O. Sadiku, M. Tembely, and S. M. Musa, "Climate-Smart Agriculture," *International Journal of Advanced Research in Computer Science and Software Engineering*, vol. 7, no. 2, February 2017, pp. 148–149.
11. R. Arshad, S. Zahoor, and M. A. Shah, "Green IoT: An investigation on energy saving practices for 2020 and beyond," *IEEE Access*, vol. 5, 2017, pp. 15667–15681.
12. S. Din, A. Ahmad, and A. Paul, "Human enabled green IoT in 5G networks," *Proceedings of the Symposium on Applied Computing*, Marrakech, Morocco, April 2017, pp. 208–213.
13. S. S. Prasad and C. Kumar, "A green and reliable Internet of things," *Communications and Network*, vol. 3, 2013, pp. 44–48.

9

Green Cloud Computing

Self-discipline is the ability to make yourself do what you should do,
when you should do it, whether you feel like it or not.

—**Elbert Hubbard**

9.1 Introduction

The effects of global warming are noticeable all over the world. Rising levels
of global warming and environmental concerns have made companies to be
aware of their carbon footprint, energy consumption, and e-waste. The major
sources of CO_2 emissions include electricity, transportation, industry, and
agriculture. It is well known that cloud networks and data centers consume
a lot of energy. The cloud computing infrastructure is not only expensive
to maintain, but also unfriendly to the environment. Today, organizations
have come to realize that going green is in their best interest and are moving
toward implementation of green cloud computing (GCC). GCC is becoming
a popular trend today with the emergence of Internet-driven services in all
areas of life [1].

Cloud computing refers to the vast economies of scale, rapid market adoption velocity, and potential revenue growth of cloud computing initiatives.
It provides computing power and resources as a service to users worldwide.
However, the growing demand of cloud infrastructure has significantly
increased the energy consumption that leads to high carbon emissions,
which is not environmentally friendly. Cloud service providers are being
asked to be responsible toward the society by reducing the environmental
impact of their business operations while keeping the desired quality of service [2]. Energy-efficient solutions are required to minimize the environmental impact of cloud computing.

In recent years, there have been two major trends in the information communication technology (ICT) industry: green computing and cloud computing.
The combination of these two trends leads to GCC. The GCC is an energy-
efficient tool and also makes business more environmentally responsible.
It utilizes resources efficiently so as to decrease the impact of IT processes
on the environment. It seeks to achieve the sustainable development of
cloud computing and reduce the possible impact of cloud systems on the

environment. GCC can produce solutions that can make the IT resources energy efficient while minimizing the operational costs. GCC is known to be a hot area for research [3].

This chapter provides an introduction to GCC. It begins by providing an overview of cloud computing and green computing. It covers several efficiency metrics and presents some economic benefits and challenges facing GCC. The last section concludes the chapter.

9.2 Overview of Cloud Computing

Cloud computing (or cloud) is one of the most popular emerging technologies today. It enables outsourcing of all IT needs such as storage, computation, and software through the Internet. It refers to the remote computing resources, usually in data centers that provide services to users over the Internet. In cloud computing, data is usually stored, retrieved, and processed in the data centers. Cloud refers to a pool of data centers on which various services are deployed through the Internet and also on which data is stored, retrieved, and processed. A typical data center has three main components: data storage, servers, and a local area network. Data centers are set up at multiple geographical locations to facilitate distributed users.

Technologies that are working behind the cloud computing include [4]: (1) virtualization, (2) service-oriented architecture (SOA), (3) grid computing, and (4) utility computing. The key feature of cloud computing is the idea of virtualization, which enables an operating system to run on several hardware deployments. Cloud computing is a superset of grid computing. Grid computing refers to a distributed architecture of a large number of computers connected to solve a complex problem. Like electricity supply, the cloud provides a new kind of "utility" that is delivered through wired or wireless networks.

Common characteristics of clouds include on-demand self-service, broad network access, resource pooling, rapid elasticity, and measured service. The key attributes that distinguish cloud computing from conventional computing are listed below [5]:

- Computation and storage functions are abstracted and offered as services
- Services are built on a massively scalable infrastructure
- Services are delivered on demand through dynamic, flexibly configurable resources
- Services are easily purchased and billed by consumption
- Resources are shared among multiple users (multi-tenancy)
- Services are accessible over the Internet

In cloud computing, the cloud providers are responsible for hardware and software management. Cloud providers include Google, Apple, Microsoft, the International Business Machines Corporation (IBM), and Yahoo; they operate clouds commercially [6]. They are rapidly deploying data centers in various locations around the world to provide users with a variety of cloud computing services.

Cloud services are offered as "metered" services where providers have an accounting model for measuring the use of the services [7]. Some consider cloud computing as the best way to increase energy efficiency and create a more sustainable environment. The Cloud Computing Interoperability Forum (CCIF) is developing a standardized cloud computing ecosystem to achieve cloud interoperability.

Cloud computing provides three service models and four deployment models. The available service models are classified as SaaS (Software-as-a-Service), PaaS (Platform-as-a-Service), and IaaS (Infrastructure-as-a-Service). Virtualization plays a crucial role in managing these services. Cloud infrastructures use virtualization techniques for sharing of resources. The cloud deployments are classified mainly into four types: Public Cloud, Private Cloud, Community Cloud, and Hybrid Cloud, as illustrated in Figure 9.1 [8]. A public cloud is characterized by public availability of the cloud services. It describes cloud computing in the traditional mainstream sense. It is a pay-as-you-go implementation for the public, while a private cloud is not made available to the public. A community cloud is utilized via a collection of organizations that have common interest. A hybrid cloud is a combination of a public cloud and a private cloud [9].

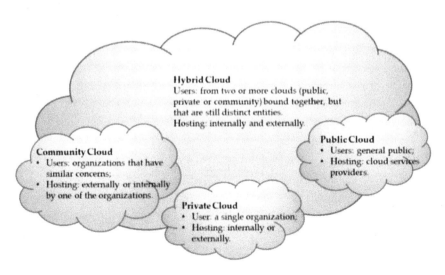

FIGURE 9.1
Types of cloud computing [8].

Cloud is an energy-hungry technology with numerous data centers running and consuming energy 24/7. The energy consumption of data centers is a bottleneck in the cloud computing technology since cloud data centers consume inordinate amounts of energy. The cloud providers deploy data centers that need energy for monitors, consoles, fans for processors, cooling system, etc. This high demand of energy tends to increase the cost and carbon emission, which reduces its efficiency. Using renewable energy sources for cloud computing is important and environmentally friendly. Implementing green computing solutions is wise not only from a moral standpoint, but also from a profit-making standpoint.

Many businesses are leveraging cloud computing benefits such as faster scale-up/scale-down of capacity, pay-as-you-go pricing, and access to cloud-based services and applications. The pay-as-you-go pricing compels the users to consume just what is needed and nothing more. With cloud computing, we can reduce the e-waste production by reducing hardware and software. We are in dire need of GCC solutions that can not only save energy, but also reduce operational costs. Some claim that cloud computing is also a green solution since it addresses two critical elements of a green IT approach: energy efficiency and resource efficiency.

9.3 Green Computing

Cloud computing and green computing are two most recently emerging areas in ICT. They are developing very fast and drastically changing the traditional way of computation. Green computing refers to the eco-friendly and environmentally responsible usage of computers and their resources. It is about reducing the environmental footprint of ICT. It is the practice of using computing resources efficiently. It is the movement toward a more environmentally sustainable computing. The goal of green computing is to reduce the use of hazardous materials and minimize factory waste. There is no easy way to green computing. We must strive to minimize greenhouse gases and waste, while increasing the effectiveness of IT, such as computers, data centers, and computer networks. A green computing activity must cover all territories: people, organizations, equipment, and networks.

Cloud computing is the latest trend in the field of green computing. It is a combination of green computing and cloud computing. It does away with the hardware servers and uses virtual servers. Thus, cloud computing is an energy-efficient technology for ICT [10]. Green computing is the future technology that supports environment, reuses consumed power, and optimizes the resources efficiently.

9.4 Green Cloud

In green cloud, green means environment-friendly, while the cloud represents the Internet service delivery model. The development of GCC is related to the evolution of green data centers since the data centers are the core of cloud computing. The goal of GCC solutions is not only to save energy but also to minimize the energy consumption. GCC provides unlimited computation, unlimited storage, and service delivery through the Internet, as conceptually illustrated in Figure 9.2 [11]. Green cloud can help consolidate workload and achieve significant energy saving for the cloud computing environment. It also guarantees the real-time performance for many performance-sensitive applications [12].

The data centers employ thousands of processors in multiprocessor servers. Like other data centers, green data centers have resources that are the software and portion of green cloud. Designing green cloud data centers will be one the solutions for IT industries [13]. Virtualization can be used to effectively reduce power consumption by data centers.

The GCC concept has an important role in reducing energy consumption in IT industries. GCC is becoming important due to the increasing concerns about environmental issues by cloud service providers. Data management and efficient infrastructure are critical to facilitate GCC. With the ever-increasing use of mobile devices, green mobile communications would be a foundation for GCC [14]. For cloud providers to provide green services, they must invest in renewable energy sources.

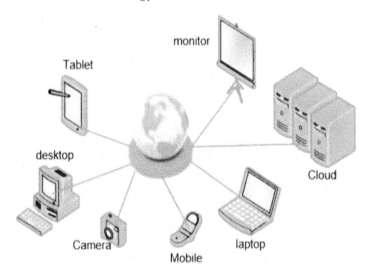

FIGURE 9.2
Green cloud computing [1].

FIGURE 9.3
Life cycle of green computing [15].

The life cycle of green computing includes analysis, specification, design, implementation and usage, and recycling and disposal phases, as shown in Figure 9.3 [15]. From the cloud provider side, each cloud layer requires becoming "green" conscious. Cloud-based services ranging from servers, Software-as-a-service, and Infrastructure-as-a-service will all contribute to decreased e-waste. For example, the SaaS providers need to model energy efficiency of their software design, implementation, and deployment. Various green scheduling and resource provisioning policies will ensure minimum energy usage.

9.5 Green Data Centers

A green data center is the foundation of GCC. A data center is a building or facility composed of networked computer systems and storage that businesses use to organize, process, store, and disseminate large amounts of data. It requires cooling and powering of servers, which results in a high carbon footprint. The creation of green, sustainable data centers has become indispensable. A green or environment-friendly data center avoids polluting the atmosphere as much as possible, and thus helps to green the surroundings. Data center resources must be managed in an energy-efficient manner to drive GCC.

To go "green," data centers must measure, analyze, and control power usage. This is challenging chip microprocessor, server, storage, and network designs to deliver higher performance within strict power and cooling constraints. Data centers can go green by saving electricity in two main areas: computing and cooling. A green or sustainable data center is a type of server facility that utilizes energy-efficient technologies. It aims to create energy-efficient data centers heading toward a greener and more sustainable environment. It is to "minimize" the overall consumption of "brown" energy (i.e., energy generated by carbon-intensive means) by data centers and maximize their use of green energy. It is designed for minimum environmental impact through the use of low-emission building materials. Green computing is one of these technologies that can help lessen carbon emissions in data centers. This requires installing energy-efficient hardware and virtualization.

Energy efficiency is regarded as a desirable outcome. One of the most important parameters to document is the amount of energy in kilowatts the data center is using. Documentation of energy usage is critical to track changes in servers, develop efficiency metrics, and use virtualization [9]. With an increasing concern over global climate change and a rapid rise in energy consumption on the data-driven market, some data centers now use renewable energy such as solar or wind power. This type of energy is termed green energy. Figure 9.4 shows the components of a solar-powered green data center [16].

Virtualization is a software-based technology that allows virtual machines (VMs) to share the same hardware. The rise of virtualization has added an important dimension to the data center infrastructure. Virtualization has allowed for more productive use of information technology equipment, resulting in much higher efficiency and lower energy. It supports the abstraction of servers, networks, and storage, allowing every computing resource to be

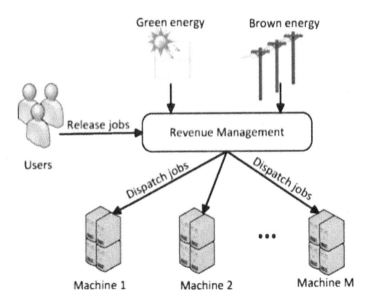

FIGURE 9.4
The components of a typical solar-powered green data center [16].

organized into pools without regard to their physical location. When coupled with the green design of new server and storage hardware, virtualization offers an effective approach to keep power and cooling costs in check [17].

9.6 Metrics

There are several efficiency metrics that illustrate the relationship between electricity consumption and CO_2 emissions. Green Grid Consortium coined a derivation to calculate the energy productivity in data centers using Power Usage Effectiveness (PUE) and its reciprocal, data center efficiency (DCE) metrics. PUE is the ratio of the total amount of power consumed by a data center to the power consumed by IT devices, such as servers, routers, storage networking devices, etc. If the value of PUE is 1, this means that the data center is 100% energy efficient. By using the most energy-efficient technologies, cloud providers can significantly improve the PUE of their data centers.

PUE is a widely adopted metric that does not offer adequate information about the energy efficiency of services. To have a comprehensive overview of the greenness of services, PUE needs to be used in conjunction with several other efficiency metrics. To avoid confusion between different actors, the quality of services should be determined in service-level agreements (SLA). Green SLAs underline the importance of having uniform metrics that illustrate energy efficiencies and CO_2 emissions. Other metrics include Energy Reuse Effectiveness (ERE) or Energy Reuse Factor (ERF), Carbon Usage Effectiveness (CUE), Technology Carbon Efficiency (TCE), Data Center Productivity, and Compute Power Efficiency (CPE) [18, 19].

9.7 Benefits and Challenges

The popularity of the cloud due to the numerous financial and nonfinancial benefits that it provides has enabled its adoption by many organizations. The benefits of implementing green cloud computing are as follows [20]: (1) reduces paper waste, (2) reduces energy consumption, (3) reduces footprint, (4) allows employees to telecommute, (5) takes advantage of government incentives, (6) obtains lower costs with high profit, (7) complies with regulation, and (8) improves resiliency. GCC is an energy-efficient tool in the workplace. It also reduces the business ecological impact. When your company uses GCC, you save energy through resource pooling. GCC solutions save energy and reduce operational costs. It makes it possible to work from anywhere at any time so that employees can work from home if needed.

Cloud computing faces several challenging issues related to security, load balancing, quality of service, standardization, and energy consumption. Perhaps the biggest challenge to GCC is related to security. Security is an important factor enabling GCC infrastructure to be deployed. Security issues include sensitive data access, privacy, data recovery, and multi-tenancy issues. Customers should be able to trust that cloud service providers will not misuse their sensitive data [18]. Energy consumption is another main obstacle to GCC. Load balancing in achieving GCC is another challenge. Load balancing is required to distribute the dynamic workload across multiple nodes and avoid overwhelming single node while other nodes are idle [21]. Another limitation is the high cost of purchase of components that are required to make the cloud computing more efficient (such as cooling equipment). The maintenance of the devices included in data centers is also a major concern.

9.8 Conclusion

GCC is regarded as a hot area for research. It is becoming important in a world with limited energy resources and an ever-increasing demand for more computational power. It is also becoming important due to the increasing concerns about environmental issues by the cloud service providers. It has a potential to be a powerful technology that can contribute to green IT and carbon emissions. Although the research on GCC is still at an early stage, it is becoming more and more popular. It is the latest trend today and is the future of cloud computing. More information about GCC can be found in the book in Reference [22, 23].

References

1. M. N. O. Sadiku, C. M. M. Kotteti, and S. M. Musa, "Green cloud computing," *International Journal of Scientific Engineering and Technology*, vol. 7, no. 10, October 2018, pp. 94–99.
2. P. N. Balasooriya, S. Wibowo, and M. Wells, "Green cloud computing and economics of the cloud: Moving towards sustainable future," *GSTF Journal on Computing*, vol. 5, no. 1, 2016, pp. 15–20.
3. Y. S. Patel, N. Mehrotra, and S. Soner, "Green cloud computing: A review on green IT areas for cloud computing environment," *Proceedings of the 1st International Conference on Futuristic Trend in Computational Analysis and Knowledge Management*, 2015, pp. 327–332.
4. Y. Goyal, M. S. Arya, and S. Nagpal, "Energy efficient hybrid policy in green cloud computing," *Proceedings of International Conference on Green Computing and Internet of Things*, October 2015, pp. 1065–1069.

5. T. A. M. Sa'ed, "Toward green and mobile cloud computing," *Proceedings of IEEE Seventh International Conference on Intelligent Computing and Information Systems,* 2015, pp. 203–209.

6. M. N. O. Sadiku, S. M. Musa, and O. D. Momoh, "Cloud computing: Opportunities and challenges," *IEEE Potentials,* vol. 33, no. 1, January–February 2014, pp. 34–36.

7. S. K. Garg and R. Buyya, "Green cloud computing and environmental sustainability," http://www.cloudbus.org/~raj/papers/Cloud-EnvSustainability2011.pdf

8. L. D. Radu, "Green cloud computing: A literature survey," *Symmetry,* vol. 9, 2017.

9. N. M. Rawai et al., "Cloud computing for green construction management," *Proceedings of Third International Conference on Intelligent System Design and Engineering Applications,* 2013, pp. 432–439.

10. M. N. O. Sadiku, N. K. Nana, and S. M. Musa, "Green computing: A primer," *Journal of Scientific and Engineering Research,* vol. 5, no. 4, 2018, pp. 247–251.

11. S. H. Alsamhi et al., "Greening Internet of things for smart everything with a green environment life: A survey and future prospects," https://arxiv.org/ftp/arxiv/papers/1805/1809.00844.pdf

12. L. Liu et al., "GreenCloud: A new architecture for green data center," *Proceedings of the 6th International Conference Industry Session on Autonomic Computing and Communications Industry Session,* Barcelona, Spain, June 2009, pp. 29–38.

13. K. Khajehei, "Green cloud and reduction of energy consumption," *Computer Engineering and Applications,* vol. 4, no. 1, February 2015, pp. 51–60.

14. F. S. Chu, K. C. Chen, and C. M. Cheng, "Toward green cloud computing," http://citeseerx.ist.psu.edu/viewdoc/download?doi=10.1.1.909.1011&rep=rep1&type=pdf

15. H. A. Akpan and R. J. Vadhanam, "Green cloud technique: A survey," *International Journal of Applied Engineering Research,* vol. 10, no. 82, 2015, pp. 116–121.

16. H. Wang et al., "On time-sensitive revenue management in green data centers," *Sustainable Computing: Informatics and Systems,* vol. 14, 2017, pp. 1–12.

17. IBM, "The green data center. More than social responsibility: A foundation for growth, economic gain and operating stability," May 2007.

18. T. Makela and S. Luukkainen, "Incentives to apply green cloud computing," *Journal of Theoretical and Applied Electronic Commerce Research,* vol. 8, no. 3, December 2013, pp. 74–86.

19. A. M. Farooqi, T. Nafis, and K. Usvub, "Comparative analysis of green cloud computing," *International Journal of Advanced Research in Computer Science,* vol. 8, no. 2, March 2017, pp. 56–60.

20. E. Tabor, "4 green benefits of cloud computing," https://www.isgtech.com/4-green-benefits-of-cloud-computing/

21. N. J. Kansal and I. Chana, "Cloud load balancing techniques: A step towards green computing," *International Journal of Computer Science Issues,* vol. 9, no. 1, January 2012, pp. 238–246.

22. A. K. Sangaiah and N. S. Kumar, *Sustainable Green Cloud Computing.* Boca Raton, FL: CRC Press, 2019.

23. K. Munir, *Cloud Computing Technologies for Green Enterprises.* Hershey, PA: IGI Global, 2018.

10

Green Communications and Networking

Nothing can stop the man with the right mental attitude from achieving his goal; nothing on earth can help the man with the wrong mental attitude.

—Thomas Jefferson

10.1 Introduction

The demand for ubiquitous wireless and Internet services has been on the rise for the past decades. This requires a considerable amount of energy, which is often underestimated.

Advances of mobile communication devices, such as smart phones, smart watches, and wearable healthcare devices, have moved us toward the era of smart society. Such devices have become an indispensable part in our daily life because they allow us to exchange information reliably from anywhere, any time. The global wireless data traffic shows no signs of slowing down [1]. However, there has been increase in the unnecessary energy consumption of the mobile communication devices. The increasing volume of transmitted data is sustained at the expense of a significant carbon footprint by the mobile communications industry. The implication of wireless network's environmental and social responsibility (energy efficiency and environmental impact) has been disregarded. Computers themselves risk becoming the "energy hogs" of the future, unless something is done. Powering over one billion personal computers, over four billion fixed and mobile telephones, and computer networks (including the Internet and wireless networks) around the world requires approximately 1.4 Petawatt-hr a year (7015140000000000000♠1.4×10^{15} W-hr) of electricity [2]. The projected carbon footprint of the mobile communications until 2020 is shown in Figure 10.1 [3].

In spite of the great progresses in optical communication and transmission, today's networks still rely heavily on electronic systems. The proliferation of information and communication technology (ICT) systems is causing energy consumption levels to reach distressing rates. There is a request of environmental protection from users and governments to reduce CO_2 emissions due to ICT. The United Nations Climate Change Conferences have been held yearly to evaluate the progress in dealing with climate change since 1995. Green communications and networking can introduce significant

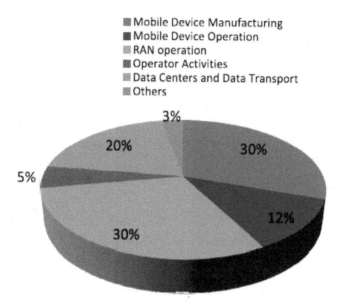

FIGURE 10.1
Carbon footprint of mobile communication (projected in 2020) [3].

reductions in energy consumption in the ICT industry. Green communication refers to communication that is sustainable, energy efficient, energy aware, and environmentally aware. It is also an environmental friendly communication. Green communication aims at balancing the resource usage and consequently saving the energy of entirely mobile and wireless networks. It is expected to address the growing cost, environmental impact, and CO_2 emission of telecommunication. With the rapid growth and evolution of communication and networking technologies, energy consumption is growing fast. Hence, green communication and networking are urgently needed [4].

This chapter provides an introduction to energy-efficient communications and networking, including wireless and wired networks. It begins by presenting some characteristics of green communication and green networking. Then, it discusses some common applications of green communication and networking. It addresses green metrics and some benefits of those technologies. The last section concludes the chapter.

10.2 Green Communications

ICT is one of the keys to a future low-carbon and sustainable society and the wireless communication networks constitute the largest share of the ICT. Communications technologies will be critical to achieve large-scale energy savings. Reduction of the greenhouse gases (GHGs) caused by the telecommunication sector is known as greening of telecommunication, which has many facets. It can be classified broadly in terms of greening of telecommunication networks, green telecommunication equipment manufacture, atmosphere friendly design of telecommunication buildings, and safe telecommunication waste disposal [5].

The term "green" refers to recycling, purchasing, and using environment-friendly products. The key idea behind the concept of green communication is to find a way on how to encourage people to change their behavior in order to increase efficient use of communication systems. Green communication aims at reducing energy cost while still maintaining quality of service (QoS) in terms of coverage needs, capacity, and user needs.

Green communications require that all blocks in a communication system (the baseband, the transmitter, the receiver, and the signal modulation) are designed for optimum efficiency. The goal of green communication is to ensure that communication systems consume less energy and have a smaller carbon footprint. Strategies for achieving this include using renewable energy, biodiesel, and solar and fuel-powered cell sites, and installing fuel catalysts and cooling units [6, 7].

Four fundamental trade-offs have been identified [8]: deployment efficiency—energy efficiency to balance the deployment cost and energy consumption in the network, spectrum efficiency—energy efficiency to balance the achievable rate and energy consumption, bandwidth—power to balance the bandwidth utilized and the power needed for transmission, and delay power—power to balance the average end-to-end service delay and average power consumed. Figure 10.2 illustrates these trade-offs [8].

Green communication satisfies the same criteria for green technology [9, 10]:

a. It minimizes the degradation of the environment.

b. It has zero or low GHG emission.

c. It promotes healthy and improved environment for all forms of life.

d. It conserves the use of energy and natural resources.

e. It promotes the use of renewable resources.

Key techniques of green communication mainly include cognitive network, network coding, and smart grid [11]:

- *Cognitive network:* This network can effectively improve the spectrum resource utilization efficiency and the network transmission performance. Cognitive radio plays a crucial role in improving the utilization efficiency of radio spectrum. It is capable of exploiting the residual bands when their licensed users (known as primary users) are not broadcasting on those frequencies, and to free up the channel as soon as the primary users want to access it. As illustrated in

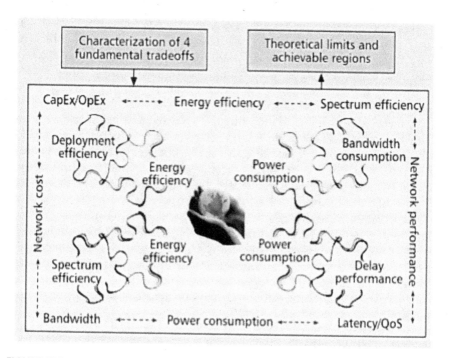

FIGURE 10.2
Fundamental trade-offs [8].

Figure 10.3, cognitive radio adds a dimension of intelligence, learning, and adaptation [12].

- *Network coding:* This involves removing redundant routes to improve the network throughput. Network coding technology saves network bandwidth and improves the link utilization.

- *Smart grid:* The main objective of the smart grid is to bring reliability, flexibility, efficiency, and robustness to the power system. The

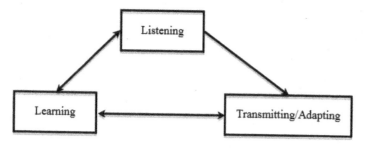

FIGURE 10.3
How cognitive radio works [12].

smart grid does this by introducing two-way data communications into the power grid. It provides the modern electricity grid with a high-speed, fully integrated, two-way communication technological framework. It facilitates measuring, monitoring, protecting, and controlling functions.

10.3 Green Networking

Green networking (also known as energy-efficient networking) refers to minimizing utilization of energy through use of energy-efficient technology, renewable energy resources, and environmental friendly consumables. It addresses unnecessary energy consumption in the two areas: wired networks and wireless networks. Traditionally, wired networks have been designed without considering energy efficiency. But reduction of unnecessary energy consumption is becoming a major concern in both wired and wireless networks. Green networking covers all components of the network, including personal computers, switches, routers, communication media, and other devices connected to the network. Energy-efficient networking targets the reduction of energy consumption by these components. Some of the goals of green networking include [13]:

i. Reduction of energy consumption.

ii. Improvement of energy efficiency.

iii. Consideration of the environmental impact of network components from design to end of use.

iv. Integration of network infrastructure and network services.

v. Making the network more intelligent.

vi. Compliance with regulatory reporting requirements.

vii. Promotion of a cultural shift in thinking about how we can reduce carbon emissions.

Green networking is the practice of selecting energy-efficient networking technologies and products, and minimizing resource use. Its practices include [14]:

1. Implementing virtualization.

2. Practicing server consolidation.

3. Upgrading older equipment for newer, more energy-efficient products.

4. Employing systems management to increase efficiency.

5. Substituting telecommuting, remote administration, and videoconferencing for travel.

Although investing in green networking may require an initial cash outlay, the products and practices involved typically save money once put in place.

Virtualization as the name implies is creating a virtual environment. Instead of running different servers separately, through virtualization all the servers can run on one to two computers [15]. Virtualization can be applied to different resources, including network links, storage devices, and software resources.

Green networking research stems from different observations on the root causes of energy waste [16]: (1) adaptive link rate, (2) interface proxying, (3) energy-aware infrastructure, and (4) energy-aware applications.

10.4 Energy-Saving Techniques

The design of energy-efficient networks is important for green communications and networking. Energy efficiency in communication systems relies mainly on the power consumption of individual components that comprise them. It is associated with reduced energy consumption. In principle, reducing the energy consumption of communication networks can be achieved in the following ways [17]:

- *Network planning:* Optimize the physical placement of resources and enable the potential to power off network equipment.

- *Equipment reengineering:* Introduce low-energy network equipment and devices.

- *Network management:* Optimize the operation of network equipment as well as network-wide protocols and mechanisms by adjusting network resources based on the users' demand.

- *Renewable energy:* Reduce the energy consumption from power grid by supplying energy from alternative sources, such as wind, micro hydro, solar, tidal, geothermal, etc., which are also called green energy.

- *Social awareness:* Educate users to avoid wasting energy.

Energy-saving techniques will increase the efficiency of the network and reduce the economic costs as well as environmental impact.

10.5 Applications

Common applications of green communication and networking include energy efficiency in wireless networks, cellular networks, mobile networks, vehicular ad hoc networks (VANETs), and smart grids. We have covered smart grids earlier.

- *Green wireless communication:* The most boisterous revolution in communication has been the rapid evolution in wireless and mobile communication. Wireless communication is the most emergent and accepted area of communication field. Green wireless communications strive for improving energy efficiency as well as reducing environmental impact. It can be achieved with the use of green handover, green codes, green electronics, green power amplification, green antennas, green manufacturing, and green base transceiver stations using renewable energy sources [18]. Other issues related to green wireless networks include green cellular base station, energy-efficient mobile terminals, green ad hoc and sensor network, green cognitive radios, electromagnetic pollution mitigation, and energy-efficient signal processing techniques [19]. Green wireless communication will provide energy-efficient communication. Green communication among mobile and wireless networks enables providing potential benefits for balancing the resource usage and saving energy. Green communication becomes the utmost important and promising research topic for future mobile and wireless networks. The next-generation wireless networks should be able to provide high-speed Internet access anywhere and anytime.
- *Cellular networks:* A cellular network is a radio network distributed over land areas known as cells, each served by at least one base station. A typical cellular network consists of three main elements: a core network that takes care of switching, base stations providing radio frequency interface, and the mobile terminals used in making voice or data connections [20]. The cellular network is the largest factor contributing to the mobile industry's energy consumption. As a result, energy efficiency in cellular networks has been a growing concern for cellular operators. Green communication technologies are widely preferred and deployed for achieving energy efficiency.
- *Green mobile communications:* It has been observed that mobile operators are among the top energy consumers. The explosive growth of mobile communications gives rise to a compelling case to reduce the electromagnetic pollution. Causes of carbon footprint in mobile communications include manufacturing of mobile devices, radio frequency (RF) transmission, charging of batteries, electricity

consumption of base stations, data centers, and data transport, operation of offices, stores, vehicle fleet, and business travel [21]. Green
mobile communications is in tune with the trend of reducing carbon
footprint and emission of GHGs [22]. To enable power saving, mobile
terminals when not in use can operate in either sleep mode or idle
mode.

- *VANETs:* VANETs provide communications required to deploy intelligent transportation, which is an emerging transportation system
that applies ICT to enhance safety and mitigate traffic congestion.
With the proliferation of electrical vehicles powered by finite batteries, any power consumption should be minimized to extend the
range of vehicles. The power consumption by VANETs is becoming
a concern and the adoption of green communications and networking is highly desirable [23].

10.6 Green Metrics

Green metrics (or energy efficiency metrics) are important in green communications and green networking. The metrics provide information that can
be used to assess and compare solutions and technologies. Energy-efficient
metrics are needed to measure the energy consumption of a network. A
number of metrics have been proposed to efficiently evaluate the effort for
achieving green communications systems. The metrics have been classified
in three main categories: component-, equipment-, and network-level metrics. Greenness metrics (such as Green Performance Indicator and Carbon
Emission Calculator) are directly involved into CO_2 emission. Component-
level metrics measure the performance of a specific component of a wireless
device such as antenna, power amplifier, power supply, etc. Equipment-level
metrics are used to investigate the energy efficiency of a given equipment
such as end user terminals or base stations. Network-level metrics assess
energy efficiency at the network level [24].

10.7 Benefits and Challenges

Adopting green communication technology helps individuals and businesses to reduce power consumption and lower the costs of operation.
Green Networking helps to reduce the carbon footprint of the ICT industry. More benefits of green communications and networking include [25]

the following: (1) reduces energy-related costs, (2) attracts new customers and increases sales, (3) may be eligible for tax incentives, (4) boosts workforce morale and innovations, (5) makes societal impact, (6) gains a competitive advantage, and (7) enters new markets with environmentally focused products.

Rising energy costs and environmental policies are some of issues that are driving the need for energy-efficient networking. One of the major barriers for deployment of energy-saving techniques in real networks is their integration in actual network protocols. The energy consumption of Internet is becoming huge due to an increase in the number of connected devices. The challenge is to make the increase of the Internet operational efficiency faster than the rate of traffic growth [17]. It is not easy to characterize the different sources of energy consumption due to lack of precise and realistic models in support of the design of protocols and algorithms. Other challenges include scalability, reliability, interoperability, and security.

10.8 Conclusion

The exponential growth of mobile and wireless systems makes it difficult to ignore their carbon footprint. Green communication and green networking are of utmost importance for the future mobile and wireless networks. Their goal is to drive the technology toward "Going Green." It is important that any emerging technology is harmonized with our mother nature. This means better fuel efficiency and better processes would mean even better things for the environment.

The need for adopting green communication and networking has been realized worldwide. Green communications are key to facilitate real smart applications. Green communications and networking will meet the demand of energy efficiency of the next-generation wired, wireless, smart-grid networks, and IoT. They are the new focus of the telecommunications industry. They have attracted significant attention from academia, industry, and government agencies due to their ability to create eco-friendly power-efficient networks. It has been observed that telecommunications applications can have a significant impact on lowering GHG emissions and power consumption. Consequently, organizations are following environment green policies in hiring, training, and running of their business.

For more information on green communication, one should consult books in References [12, 17, 26–32] and other books available on Amazon.com. One should also consult the journal exclusively devoted to it: *IEEE Transactions on Green Communications and Networking*.

_____

References

1. M. N. O. Sadiku, M. Tembely, and S. M. Musa, "Green communications: An introduction," *Journal of Multidisciplinary Engineering Science and Technology*, vol. 3, no. 7, July 2016, pp. 5216–5217.

2. "Green Comm Challenge," *Wikipedia*, the free encyclopedia, https://en.wikipedia.org/wiki/Green_Comm_Challenge

3. P. Gandotra, R. K. Jha, and S. Jain, "Green communication in next generation cellular networks: A survey," *IEEE Access*, vol. 5, 2017, pp. 11727–11758.

4. A. Abrol and R. K. Jha, "Power optimization in 5G Networks: A step towards green communication," *IEEE Access*, vol. 4, 2016, pp. 1355–1374.

5. L. K. Chhaya, "Green wireless communication," *Journal of Telecommunications System & Management*, vol. 1, no. 2, 2012.

6. M. N. O. Sadiku, A. E. Shadare, and S. M. Musa, "Green communication," *International Journal of Trend in Research and Development*, vol. 5, no. 4, July–August 2018, pp. 211–212.

7. M. Kocaoglu, D. Malak, and O. B. Akan, "Fundamentals of green communications and computing: Modeling and simulation," *Computer*, vol. 45, September 2012, pp. 40–46.

8. Y. Chen et al., "Fundamental trade-offs on green wireless networks," *IEEE Communications Magazine*, vol. 49, June 2011, pp. 30–37.

9. M. Bhardwaj and Neelam, "The advantages and disadvantages of green technology," *Journal of Basic and Applied Engineering Research*, vol. 2, no. 22, October–December 2015, pp. 1957–1960.

10. M. N. O. Sadiku et al., "Green Technology," *International Journal of Trend in Scientific Research and Development*, vol. 3, no. 1, November–December 2018, pp. 1137–1139.

11. F. Chuan and L. Anqing, "Key techniques in green communication," *Proceedings of the International Conference on Consumer Electronics, Communications and Networks*, XianNing, China, April 2011, pp. 1360–1363.

12. F. Krief (ed.), *Green Networking*. Wiley-ISTE, 2013, p. 118.

13. N. Chilamkurti, S. Zeadally, and F. Mentiplay, "Green networking for major components of information communication technology systems," *EURASIP Journal on Wireless Communications and Networking*, 2009, doi: 10.1155/2009/656785.

14. J. Marsh, "What is green communication?" https://www.quora.com/What-is-green-communication

15. D. E. Chowdary, N. R. Vaishnav, and G. Apoorva, "Green networking using a combination of network virtualization and adaptive link rate," *Proceedings of IEEE International Conference On Recent Trends In Electronics Information Communication Technology*, India, May 20–21, 2016, pp. 1550–1553.

16. P. Bianzino et al., "A survey of green networking research," *IEEE Communications Surveys and Tutorials*, vol. 14, no. 1, 2012, pp. 3–20.

17. K. Samdanis et al., *Green Communications: Principles, Concepts and Practice*. Chichester, UK: John Wiley & Sons, 2015, p. 10.

18. L. K. Chhaya, "Green wireless communication," *Journal of Telecommunications System & Management*, vol. 1, no. 3, 2012.

19. I. U. Ramírez and N. A. B. Tello, "A survey of challenges in green wireless communications research," *Proceedings of International Conference on Mechatronics, Electronics and Automotive Engineering,* Cuernavaca, Mexico, November 2014, pp. 197–200.
20. S. A. Waje and S. B. Rahane, "A survey of green wireless communications," *International Journal of Electronics, Communication & Instrumentation Engineering Research and Development,* vol. 3, no. 2, June 2013, pp. 25–310.
21. J. Thirumaran and S. Dhinakaran, "Green communications and networking systems – A challenge to current communications and protocols," *International Journal of Scientific Engineering and Research,* 2014, pp. 21–24.
22. P. Venkatapathy, I. N. Jena, and A. Jandhyal, "Electromagnetic pollution index– A key attribute of green mobile communications," *Proceedings of IEEE Green Technologies Conference,* Tulsa, OK, April 2012.
23. J. Toutouh and E. Alba, "An efficient routing protocol for green communications in vehicular ad-hoc networks," *Proceedings of the 13th Annual Conference Companion on Genetic and Evolutionary Computation,* Dublin, Ireland, July 2011, pp. 719–7210.
24. R. Mahapatra et al., "Energy efficiency tradeoff mechanism towards wireless green communication: A survey," *IEEE Communications Surveys & Tutorials,* vol. 18, no. 1, Frist Quarter 2016, pp. 686–705.
25. "Five benefits of embracing sustainability and green manufacturing," https://www.nist.gov/blogs/manufacturing-innovation-blog/five-benefits-embracing-sustainability-and-green-manufacturing
26. S. Mumtaz and J. Rodriquez (eds.), *Green Communication in 4G Wireless Systems.* Aalborg, Denmark: Rivers Publishers, 2013.
27. F. R. Yu, X. Zhang, and V. C. M. Leung (eds.), *Green Communications and Networking.* Boca Raton, FL: CRC Press, 2013.
28. N. Kaabouch and W. C. Hu, *Energy-Aware Systems and Networking for Sustainable Initiatives.* IGI Global, 2012.
29. E. Hossain, V. K. Bhargava, and G. P. Fettweis (eds.), *Green Radio Communications Networks.* Cambridge, UK: Cambridge University Press, 2012.
30. S. Murugesan and G. R. Gangadharan (eds.), *Harnessing Green It: Principles and Practices.* Hoboken, NJ: John Wiley & Sons, 2012.
31. J. Wu, S. Rangan, and H. Zhang (eds.), *Green Communications: Theoretical Fundamentals, Slgorithms and Applications.* Boca Raton, FL: CRC Press, 2013.
32. S. Khan and J. L. Mauri (eds.), *Green Networking and Communications: ICT for Sustainability.* Boca Raton, FL: CRC Press, 2013.

11

Green Growth and Economics

> The pessimist sees difficulty in every opportunity; an optimist sees the opportunity in every difficulty.
>
> —**Winston Churchill**

11.1 Introduction

Global warming, financial crisis, rapid depletion of natural resources, environmental degradation, and globalized marketplace are major problems facing the world today and have their root in the dominant economic system. The current global economic growth pattern is environmentally unsustainable. Hence, the world is in dear need of a new economics or an economic system overhaul. The advocates of green economics argue that economies are social structures that should respond to social and environmental priorities [1].

The modern economy was created due to innovation and thrives on it. The economy encourages new ways of doing things and the invention of new products. The term "green growth" and its sister concepts, "green economy" (GE) and "sustainable development," have attracted attention worldwide in recent years. GE and green growth are related topics. Both GE and green growth aim at identifying possible ways of improving the results of economic activities. They both stimulate international attention for transforming our present non-sustainable economic structure in the direction consistent with the objectives of sustainable development. Green growth and GE are basically subsets of sustainable development. To achieve green, sustainable development, production efficiency, energy efficiency, renewable energy, new material, and new technology are required. Sustainable development provides an important context for both green growth and GE.

Green growth means fostering economic growth and development while simultaneously combating climate change and preventing costly environmental degradation and the inefficient use of resources. Green growth is a means of achieving sustainable development. The advocates of green growth argue that green and growth can go hand in hand.

GE refers to an economy that aims at reducing environmental risks and ecological scarcities. It is one in which economic growth and environmental responsibility work together to support progress and social development.

It is an economy that results in improved human well-being and social equity. Green growth describes a path of economic growth that utilizes natural resources in a sustainable manner. Governments and international agencies embrace green growth as a strategy to help them address social equity and deliver both economic and environmental gains.

To be green, an economy must be both efficient and fair. The Kyoto Protocol in 1998 demanded every nation to control the emission of carbon dioxide. The United Nations considers GE as a means for achieving sustainable development and eradicating poverty worldwide. The green economics is the green industry that includes green energy, green agriculture, green tourism, green manufacturing, green food, green growth, and green services. Since green technologies are important component of green economics, R&D in green technology is a prerequisite of green economics [2].

A GE can only be achieved through the commitment to multiple sectors. It must coexist with other sustainable development concepts as illustrated in Figure 11.1 [3]. Religious leaders, business leaders, governments, and trade unionists have hailed green stimulus as a cure for the world economy's ills. A GE is low-carbon, resource efficient, and socially inclusive. Green economics reincorporates into economics moral concerns, social and environmental justice, inclusiveness, equity, and access.

This chapter provides an introduction to green growth and green economics. It begins with addressing green growth. It then presents mainstream economics and addressing the importance of green economics. It covers the objectives and principles of green economics. It covers some economic sectors that need greening. It addresses the main benefits of adopting a GE and the challenges. The last section concludes the chapter.

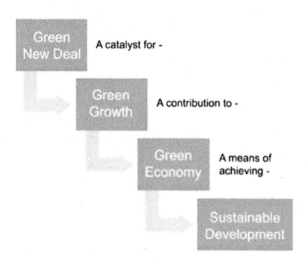

FIGURE 11.1
The hierarchy of green economy concepts [3].

11.2 Green Growth

Economic performance is usually measured by growth in GDP per capita. While development is a societal goal, economic growth is a way to achieve it. Economic development has improved the living conditions of people worldwide. Since the industrial revolution, economic growth has witnessed an amazing success. However, economic growth has come at the expense of the environment. In view of growing global competition for resources, the old growth model will be self-defeating.

The need for greening economic growth has become an international concern and an important part of national development. The need comes from the understanding of the economic costs associated with climate change and ecosystem degradation. Greening the economy is a new strategy for social transformation, simultaneously reducing environmental pressures, promoting economic growth, and enhancing social well-being.

Green growth describes a path of economic growth, that utilizes natural resources in a sustainable manner. Governments and international agencies embrace green growth as a strategy to help them address social equity and deliver both economic and environmental gains.

The concept of green growth has its origins in the Asia and Pacific Region. At a ministerial conference held in March 2005 in Seoul, 52 governments and other stakeholders from Asia and the Pacific agreed to move beyond the sustainable development rhetoric and pursue a path of "green growth." Since then, policy makers and practitioners have largely embraced green growth. The main goal of green growth is to establish incentives that will increase well-being by improving resource management and boosting productivity [4].

Green growth describes a new path of economic growth and development, that uses natural resources in a sustainable manner. It is about making growth processes resource-efficient, cleaner, and more resilient without necessarily slowing them. It is related to sister terms such as "green economy" and "sustainable development." Figure 11.2 shows the relationship between green growth, GE, and sustainable development [5]. Growth, often measured with a metric such as GDP, is now recognized as a critical driver of poverty reduction and improvements in social indicators.

11.2.1 Green Growth Strategies

The idea of green growth is generating a variety of political positions, from enthusiastic to cautious. National and international efforts to promote green growth have been intensifying in recent years. Greening the growth path of an economy depends on taking some growth strategies: green policies, green innovation, and GE.

- *Green policies:* Without green policies in place, the continuation of business-as-usual economic growth and development will have serious

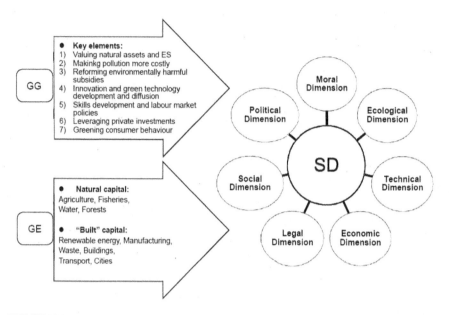

FIGURE 11.2
Relationship between green growth (GG), green economy (GE), and sustainable development (SD) [5].

impacts on natural resources. Green policies are designed to deliver environmental benefits. Green growth policies are a critical part of such efforts and an essential part of implementing sustainable development. The policies should focus on what happens over the next 5–10 years. Developing coherent policies for development can play a crucial role in creating an enabling environment for green growth. In the United States, President Barack Obama took several steps toward green growth and developed policies that would help shape the nation's GE.

- *Green innovation:* Policies to foster green innovation, as measured by patent counts, need to be adjusted to national situation. These require focusing the national public R&D effort more on fostering green innovation and using the opportunities offered by public procurement to strengthen and improve the markets for green products, thereby fostering innovation. Successful innovations are more likely to take place in fast-growing economies. Also, support to developing countries for science and development in green technologies will help increase innovation and technology advancements. Green growth policies pursue a variety of goals simultaneously.

- *Green economy:* The term GE was first coined in 1989 by a group of leading UK environmental economists. The major goal of GE is economic transformation to foster improvement of social welfare and justice and fostering the revival of an impaired global economy by combating poverty. More will be said about GE later.

Green growth strategies will be different for different countries, but all countries have opportunities to make their growth greener. Green growth will only be feasible if the broad governance and policy environment is conducive and stakeholders are confident. This will motivate businesses and consumers to undertake more environment-friendly activities.

Monitoring and measuring progress toward green growth requires some indicators. Examples of green growth/GE composite indicators include the following [6]: (1) The Global Green Economy Index, (2) The Green Economy Benchmark Index, (3) The Low Carbon Competitiveness Index (LCCI), and (4) The Climate Change Performance Index (CCPI).

11.2.2 Benefits and Challenges

Green growth is the only way to sustain economic growth and development over the long-term. It is vital to secure a brighter, more sustainable future of developing nations. It can contribute to job creation, economic prosperity, poverty reduction, and social equity. Other benefits of green growth include [7]:

1. Increased and more equitably distributed GDP—production of conventional goods and services.
2. Increased production of unpriced ecosystem services (or their reduction prevented).
3. Economic diversification, that is, improved management of economic risks.
4. Innovation, access, and uptake of green technologies, that is, improved market confidence.
5. Increased productivity and efficiency of natural resource use.
6. Natural capital used within ecological limits.
7. Other types of capital increased through use of nonrenewable natural capital.
8. Reduced adverse environmental impact and improved natural hazard/risk management.
9. Increased livelihood opportunities, income, and/or quality of life, notably of the poor.
10. Decent jobs that benefit poor people created and sustained.
11. Enhanced social, human, and knowledge capital.
12. Reduced inequality.

Green, sustainable growth comes with challenges. Achieving a GE overnight is not feasible and the costs of greening growth will depend on national ambition. Environmental policies affect relative prices and therefore change the structure of demand. The up-front capital requirements are high [8].

11.3 Mainstream Economics

Economics deals with the production and distribution of wealth, human well-being, and welfare. Mainstream economics is too bound up with concerns about profit, economic growth, and the perspective of the owners of production. Neoclassical economics has misused a narrow interpretation of Darwinism, justify capitalism, and advance the power of the fittest to preserve inequalities. It promotes supremacy of unadjusted market solutions at the expense of the needs of the people. The economic system perpetuates poverty, inequality, and social injustice. It drives today's unsustainable forms of globalization. Such an economic system is an instrument for social control, capitalism, and perpetuation of poverty [9]. It is patriarchal to its core, accounting for the rise of feminist economics.

GE differs from mainstream or conventional economic concepts of rights and the allocation of resources, which assume self-interested individualism and market competition. A comparison between mainstream and green economics is shown in Figure 11.3 [10]. Mainstream economics exploits and oppresses certain groups and over-rewards others. We need therefore to rethink the whole premise of our mainstream economics that is running into limits of expansion. Green economics offers a better alternative to mainstream economics. It demands that economic development is decoupled from the use of resources and environmental degradation. It requires the creation of positive alternatives in every sector of the economy (e.g., industry, finance, energy, governance, mobility, agriculture, and tourism). Different sector may require different economic measures.

11.4 Importance of Green Economics

The GE is an emergent approach to sustainable development that was launched in June 2012 at Rio by the United Nations Environment Programme (UNEP). It has become one of the most popular terms in global environmental forums.

Global warming has been discussed for long with scientific and political opinions varying widely. Green economics, also known as environmental economics, is the economic analysis and evaluation of solutions that reduce carbon emissions. It incorporates the difference, diversity, equity, and inclusiveness within its concepts of society. It brings to economics the core drivers of ecology, equity, and social and environmental justice. It may be seen as a tool for achieving sustainable development. It is a fundamental transformation that needs to be policy-driven. Government policy makers, companies, and organizations need comprehensive green economics to guide them in combating global warning and making decisions [11].

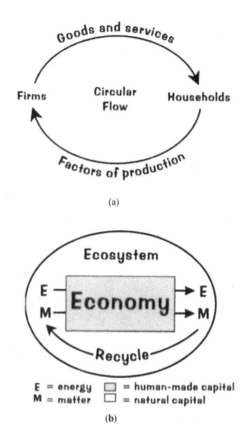

FIGURE 11.3
(a) Mainstream economy and (b) Green economy [10].

11.5 Objectives of Green Economics

There are three main objectives of green economics [12]. First, green economics creates economic conditions where social and environmental justice thrives and benefits all people. Second, green economics seeks to achieve an ambitious aim: to reform mainstream economics into a discipline which no longer supports or accepts that only a small minority can be wealthy. Green economics is providing that broad "out of the box" thinking to counteract this narrow thinking. Third, green economics goes beyond reforming mainstream economics and establishes the new discipline of green economics in order to provide the means for all people everywhere to participate in the economy with equal power, equal rights, and with equal access. Green economics attempts to create a new discipline that works for the benefit of all the people everywhere, always within the limits of nature. It reincorporates wisdom and political economy into economic problem-solving.

These objectives are achieved through the commitment and actions of multiple sectors and stakeholders in society, including government, businesses, and individuals. Participation of stakeholders is crucial to the achievement of the objectives since that provides an opportunity to understand the needs of local communities. Green economics emphasizes the importance of the local economy, local production for local needs, reusing, reducing, repairing, and possibly recycling. It advocates that it is impossible to expand forever into a finite space.

11.6 Green Economics Principles

The term GE was coined in 1989 in a pioneering report for the government of the United Kingdom by a group of leading environmental economists. Karl Burkart defines a GE as based on six main sectors [13]: (1) renewable energy, (2) green buildings, (3) sustainable transport, (4) water management, (5) waste management, and (6) land management. GEs require green energy generation based on renewable energy (including hydropower, biofuels, and biomass) to replace fossil fuels. The renewable energies can be used to help power green buildings and sustainable transport and meet the obligations and responsibilities for manufacturing. Water and waste management are being done by water purification and recycling. In nature there is no waste, as every process output is an input for some other process. GE encourages efficient use of land, including forestry, agriculture, and urbanization. It is one of the important tools for achieving sustainable development. The role of work and employment is fundamental to the success of the GE [14].

The GE can also be viewed as a set of common principles that generally include [15]:

1. The GE is a means for achieving sustainable development.
2. The GE should create decent work and green jobs.
3. The GE is resource and energy efficient.
4. The GE respects planetary boundaries or ecological limits or scarcity.
5. The GE uses integrated decision making.
6. The GE measures progress beyond GDP using appropriate indicators/metrics.
7. The GE is equitable, fair, and just—between and within countries and between generations.
8. The GE protects biodiversity and ecosystems.
9. The GE delivers poverty reduction, well-being, livelihoods, social protection, and access to essential services.

10. The GE improves governance and the rule of law. It is inclusive, democratic, participatory, accountable, transparent, and stable.
11. The GE internalizes externalities.

Green economics is one of the most holistic and multidisciplinary economics. It operates on the principle that the needs of people and natural systems must be simultaneously satisfied.

11.7 Economic Sectors Needing Greening

There is a need to fit with current macroeconomics through the creation of green jobs, poverty eradication, and growth in key sectors. The major economic sectors that need greening and green growth are building, water, waste, agriculture, energy, industry, transport, and tourism [4].

- *Buildings:* Buildings provide living and working spaces to citizens. They also have adverse impact on environmental health and the occupants' health and well-being. Green buildings are designed to mitigate these negative effects. Although most green buildings cost 0–4% more than conventional buildings, energy savings alone make green buildings cost-effective.
- *Agriculture:* The food and agricultural sectors can cause harm to environment. Due to its high exposure to climate change and its importance for the livelihoods of the majority of the population, agriculture has emerged as a priority for green growth. Greening the agriculture is important for food and nutrition requirements of future generations to be met. Green growth in agriculture sector means providing enough food to feed everyone and reducing waste in the food supply chain.
- *Energy:* Energy is crucial to economic activity and is an essential component of green growth. The importance of the energy supply to economic development is undoubtedly crucial. A major transformation in the energy sector is required because current energy system is highly dependent on fossil fuels, which contribute to the global greenhouse gas emissions. A major transformation is required in the way we produce, deliver, and consume energy. Green growth is most sustainable when it maximizes using resources that are locally available. It is perceived as shifting the sources of energy from fossil fuel to renewable energy, such as wind and solar power.
- *Manufacturing:* Manufacturing industries consume a large amount of energy and other natural resources and causes serious

environmental pollutants. They have the potential to become a driving force for realizing a sustainable society. Understanding the greening level of manufacturing industry is important to promote its green growth.

- *Business:* Advocates of green growth naturally call on businesses to adopt green policies. Greening the business sector is an opportunity to promote green investments and social entrepreneurship. For example, the bank bolsters its green credentials by increasing loans for eco-friendly farming methods. Promoting green investment in targeted areas can support economic growth. Tourism faces significant sustainability-related challenges.

- *Transportation:* Transport or mobility is a major challenge faced by developed and developing countries. The introduction of sustainable transportation systems in big cities will require strong commitment from the government.

- *Cities:* Cities experiencing rapid growth should have green growth as a top priority. Green growth in cities fostering economic growth and development through urban activities reduce negative impact on consumption of natural resources. It can foster the energy efficiency of buildings.

11.8 Benefits of Green Economy

Investment in green technology brings costs as well as benefits. The concept of green economics is powerful and influences developments in policy and politics. It has emerged as an important policy framework and practice for sustainable development in both developed and developing nations.

One of the main benefits of adopting a GE is its potential to alleviate the environmental impact caused by pollution. A GE also has a great potential to lead to economic growth, increased productivity, reduced energy bills, and lower potable water consumption. GE celebrates difference, diversity, social and environmental justice, inclusiveness, equity, fairness, and access within its concepts of society and community. The GE is regarded as a potential solution to the multiple challenges from climate change, biodiversity loss, and resource scarcity to social injustice and financial instability. It can help developing nations attain economic and social gains and reduce poverty and social inequality. Green economics helps with creating jobs, renewable energy, agriculture, manufacturing, water, and waste. Green economists believe that each person has enough to live on without exploiting other people. They promote the need to integrate all people everywhere and acknowledge the role each person plays in the economy. GE policies can help developing nations attain economic and social gains on several fronts, including creating job opportunities, sustainable agriculture, and food security.

11.9 Challenges of Green Economy

GE comes with challenges. The lack of agreement on the definition of GE creates a serious dilemma for environmental communicators and green economists. The possible misinterpretations of the concept can hamper its use in politics. Some critics are quick to point out that one cannot protect the environment without hurting some sectors of the economy. Transitioning to a GE will not be an easy process because many countries lack technology. GE requires large up-front or initial costs and long-term financing. It is a gradual process, not free of risks or costs. Few developing countries have suitable banking sectors. Progress toward a GE is hindered by insufficient financing, a limited use of economic instruments, or political emphasis on other issues.

Another challenge is cost recovery. So, we should find ways to ease cost recovery, while keeping services affordable for low-income families. Customers complain that green products are more expensive than their energy-wasting counterparts. Economics being taught in today's classroom and textbooks is outdated and needs radical transformation. Many traditional or mainstream economic ideas are not appropriate to 21st-century economies. There is urgent need for metrics that go beyond GDP for evaluating the impact of GE.

These challenges can be met with patient and careful planning of GE initiatives as well as taking advantage of what is learned from past experiences [16].

11.10 Conclusion

Green growth, a new operating strategy, is regarded as a practical, vital tool for achieving sustainable development and providing new economic opportunities. It is an effort to reconcile various aspects of economic, environmental, and social values. It is about living in harmony with nature. It is not only theoretically possible but economically achievable. Green growth has the potential to address economic and environmental challenges.

GE offers a potential solution to ecological sustainability, social justice, and lasting prosperity. The concept has gained popularity among national and international policy makers. Governments around the world are seeking ways to shape "green economy" into meaningful policies. They are developing strategies for greening their economy. The need for green economics to help determine the most economically alternatives is very urgent and timely.

The transition from mainstream economies to green economies can be implemented by cases and experiments in different industrial and social sectors. More information about green growth and green economics can be found in books in References [4, 15, 17–24] and the journal exclusively devoted to them: *The International Journal of Green Economics*.

References

1. M. N. O. Sadiku, A. E. Shadare, and S. M. Musa, "Green economics," *International Journal of Trend in Research and Development*, vol. 5, no. 6, November–December 2018, pp. 384–385.
2. W. Weiwei and Q. Lisheng, "Multiple carriers for green economy in China," *Proceedings of International Conference on Materials for Renewable Energy & Environment*, Shanghai, China, May 2011.
3. L. Georgeson, M. Maslin, and M. Poessinouw, "The global green economy: a review of concepts, definitions, measurement methodologies and their interactions," *Geography and Environment*, vol. 4, no. 1, 2017, e00036.
4. R. Kanianska, *Green Growth and Green Economy*. Banská Bystrica: Belianum, 2017.
5. A. Kasztelan, "Green growth, green economy and sustainable development: Terminological and relational discourse," *Prague Economic Papers*, vol. 26, no. 4, 2017, pp. 487–499.
6. G. Kararac et al., "Reflections on the green growth index for developing countries: A focus of selected African countries," *Development Policy Review*, vol. 36, 2017, pp. O432–O454.
7. Organization for Economic Co-operation and Development (OECD), "Green growth and developing countries: A summary for policy makers," June 2012, https://www.oecd.org/dac/50526354.pdf
8. *Inclusive Green Growth: The Pathway to Sustainable Development*. Washington, DC: World Bank, 2012.
9. M. Kennet and V. Heinemann, "Green economics: Setting the scene. Aims, context, and philosophical underpinning of the distinctive new solutions offered by green economics," *International Journal of Green Economics*, vol. 1, nos. 1/2, 2006, pp. 68–102.
10. K. Shears, "Green economics: Counting the earth in," *Green Teacher*, vol. 65, Summer, 2001, pp. 32–35.
11. D. N. Merino, "Some observations on green economics," *Engineering Management Journal*, vol. 22, no. 3, September 2010.
12. M. Kennet, "Editorial: Progress in green economics—Ontology, concepts and philosophy. Civilisation and the lost factor of reality in social and environmental justice," *International Journal of Green Economics*, vol. 1, nos. 3/4, 2007, pp. 225–249.

13. "Green economy," *Wikipedia*, the free encyclopedia https://en.wikipedia.org/wiki/Green_economy

14. M. N. O. Sadiku, S. Alam, and S.M. Musa, "Green Growth," *SSRG International Journal of Economics Management Studies*, vol. 6, no. 2, February 2019, pp. 82–84.

15. "Greeneconomy,"https://www.eea.europa.eu/publications/europes-environment-aoa/chapter3.xhtml

16. K. Pitkanen et al., "What can be learned from practical cases of green economy? Studies from five European countries," *Journal of Cleaner Production*, vol. 139, 2016, pp. 666–676.

17. P. Ekins, *Economic Growth and Environmental Sustainability: The Prospects for Green Growth*. London: Routledge, 1999.

18. International Institute for Sustainable Development, *Trade and Green Economy: A Handbook*, 3rd ed. Geneva, Switzerland: International Institute for Sustainable Development, 2014.

19. R. Hahnel, *Green Economics: Confronting the Ecological Crisis*. Armonk, NY: M. E. Sharpe, 2011.

20. J. W. Smith, *Economic Democracy: A Grand Strategy for World Peace and Prosperity; Green Economics for Sustainable Development*, 2nd ed. London: Institute for Economic Democracy, 2008.

21. D. Wall, *Babylon and Beyond: The Economics of Anti-capitalist, Anti-globalist and Radical Green Movements*. London: Pluto Press and the Green Economics Institute, 2005.

22. T. Panayotou, *Green Markets: The Economics of Sustainable Development*. San Francisco, CA: ICS Press for the International Center for Economic Growth and the Harvard Institute for International Development, 1993.

23. P. Burkett, *Marxism and Ecological Economics: Toward a Red and Green Political Economy*. Boston: Brill, 2006.

24. M. S. Cato, *Green Economics: An Introduction to Theory, Policy and Practice*. London: Imprint Routledge, 2008.

12

Green Business

Persistence. Nothing in the world can take the place of persistence. Talent will not; nothing is more common than unsuccessful men with talent. Genius will not; unrewarded genius is almost a proverb. Education will not; the world is full of educated derelicts. Persistence and determination alone are omnipotent. The slogan, "Press on," has solved and always will solve the problems of the human race.

—Calvin Coolidge

12.1 Introduction

Today there is increasing global awareness of the threat posed by climate change. Also, global warming is a challenging problem that the international community must resolve collectively. In their operations, organizations have contributed to environmental degradation due to their resource consumption, greenhouse emissions, and wastage. To reduce their impact on the natural environment, organizations must design and implement environmentally sustainable processes. There is a global consensus on the need to reduce our collective carbon footprint and reshape how we use our resources. The globalization of businesses has placed a burden on businesses to be responsible. The impact of the green has led enterprises to be more aware of the importance of sustainability, which means non-declining natural wealth. Environmental concerns have become a top issue in the business agenda. In order to eliminate the environmental pollution, the concepts of environmental management, such as green business, green management, green marketing, green growth, green production, and green innovation, etc., are now being addressed.

Green business (or sustainable business) refers to meeting customer's needs without causing environmental and social problems. It should be seen as closely related to the evolution of the green economy. Green business practices are used to reduce a company's environmental harm and achieve ecological sustainability. There are strong reasons to build greener businesses for tomorrow. Green business is an enterprise that has minimal negative impact on the environment. It is making business operations more environmentally friendly (or "green") since a green environment is a social as well as business issue. It is a low-risk, effective way of doing business.

Green businesses play a major role by utilizing renewable energy and employing green labor forces to provide green energy services and goods [1].

This chapter provides a brief introduction to green business. It begins by familiarizing the readers with conventional or traditional business and the characteristics of green business. It covers green business strategies and what it takes to build a green business. It presents the benefits and challenges of green business. The last section concludes the chapter.

12.2 Conventional Business

A business company seeks to generate profit by selling a service or a product. Its activities may include manufacturing, distribution, retailing, or franchising. In managing their operations, business companies have traditionally focused on economic imperatives in terms of time, cost, efficiency, competition, and quality. A conventional or traditional business may be a local store that offers its services or products to the customers in its neighborhood. It is a setup where customers will have to visit the store physically to buy the products. A conventional business has the following characteristics [2]:

- *Cost of infrastructure:* This is usually very high. Renting and buying office is always expensive in offline business.
- *Cost for employment:* You need to hire staff for sales, accounts, management, and security.
- *Huge investment in inventory:* Every business needs to maintain stock of the products they are dealing with.
- *Locality limitation:* The business needs to be located centrally so that the customers can visit easily.
- *Time limitation:* The business only operates at certain time, most likely not 24/7.

There are basically five key steps in each business life cycle: inputs, process, outputs, environment externalities, and marketing [3]. Businesses depend on the environment to access different inputs, such as land, energy, or water.

A business company can be divided into five elements [4]. These elements should be greened if the company wishes to achieve ecological sustainability.

1. *Mission:* The mission is typically a brief statement that states the company's central reason or purpose for existing. Companies that desire to be green should incorporate their ecological values into their mission.
2. *Employees:* No green mission can succeed without employees who are committed, informed, and proactive. Employees need to commit

to greening, be environmentally aware, be ecologically knowledge-
able, and receive proper training.

3. *Operations:* Operations involve many company activities, all of which
affect environmental health. These include accounting, marketing,
procurement, hiring, public relations, printing, fleet, and recycling.

4. *Facilities and sites:* Most facility and site activities, such as agriculture
and deforestation, do not consider the environmental impact. For
example, buildings, flooring products, furniture, and construction
materials can cause harm throughout their life cycles.

5. *Products and/or services:* Green products and services do not harm their
manufacturers, users, or the environment throughout their entire
lives. More enterprises are bringing green products to the market and
implementing eco-friendly activities. Some companies replace less
green products with greener one as a strategy toward sustainability.

Conventional business process focuses on the optimization of cost, quality,
time, and flexibility, while green business process considers the environmental sustainability, and the trade-off between them, as illustrated in
Figure 12.1 [5]. Current business practices have caused considerable environmental harm.

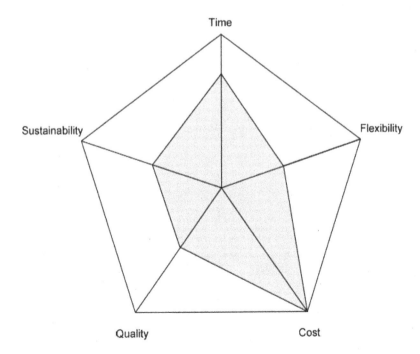

FIGURE 12.1
Recognizes sustainability as an important emergent dimension [5].

12.3 Characteristics of Green Business

Green business is part of an explosively growing economic sector that also includes green pricing, green purchase, green jobs, green IT, green management, green growth, green accounting, green innovation, green products, green customers, green factory, green manufacturing, green data centers, green infrastructure, green energy, green practices, green policies, green building, green solutions, green space, green team, green tech, and green supply chain. This clearly shows that green business touches everything. Green business is about good decision-making, risk management, employee engagement and development, capital investments, and marketing.

In general, business is regarded as green if it meets following four criteria [6]:

1. It incorporates principles of sustainability into each of its business decisions.
2. It supplies environmentally friendly products (or services) that replace nongreen products (or services).
3. It is greener than traditional competition.
4. It has made an enduring commitment to environmental principles in its business operations.

A business needs to learn how to make products green and marketable. Green products are those that save energy, avoid unnecessary waste, are recyclable, reusable, nontoxic, and not harmful to the environment. Manufacturing greener products involves including environmentally friendly materials and processing them in an ecofriendly manner [7].

12.4 Green Strategies

The aim of green business is to introduce product, processes, and services with low carbon. Green strategies are designed to achieve the aim by addressing the following three problem areas [8, 9]:

- *Process awareness:* Processes must cause minimal harm to the environment.
- *Service awareness:* Green consumerism is purchasing or nonpurchasing decisions made by consumers based on their environmental concern.

- *People awareness:* People awareness supports the achievement of strategies. Green marketing is effectively used to spread the information about green business to consumers. It refers to efforts made by the organizations to design, promote, and distribute eco-friendly products.

Green business strategies should be embraced by employees, consumers, and other stakeholders. Failure to address environmental impact of strategic business decisions may affect the profitability and competitiveness of the company.

Almost all businesses need greening. Typical ones include manufacturer, utilities, transportation, agriculture, construction industry, chemical industry, electronic industry, and food industry. They green their businesses for various reasons including competitiveness, corporate social responsibility, corporate profitability, customer satisfaction, external stakeholder pressure, and government pressure/regulation, as illustrated in Figure 12.2 [10]. Greening your business is surely a win-win situation.

In 2012, the international community came together again to develop eight Millennium Development Goals [11]:

1. Eradicate extreme poverty and hunger.
2. Achieve universal primary education.
3. Promote gender equality and empower women.
4. Reduce child mortality.
5. Improve material health.
6. Combat HIV/AIDS, malaria, and other diseases.
7. Ensure environmental sustainability.
8. Develop a Global Partnership for Development.

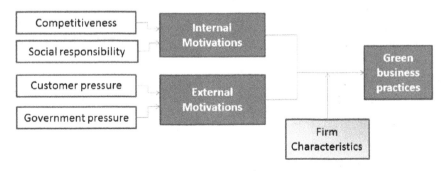

FIGURE 12.2
Motivations for green business practices [10].

The contribution of the business community toward the achievement of these goals has been through partnerships, aid donations, and other philanthropic activities.

12.5 Building a Green Business

Businesses take a variety of green initiatives. Common examples include the act of "going paperless," greenwashing, charitable donations, refurbishing used products, eliminating waste, choosing nontoxic raw materials, and supporting renewable energy. Irresponsible and indifferent attitude of some enterprises toward becoming "greener" has led to consumer boycotts and lawsuits.

There are various practices a business can use in shifting to a green behavior. Some of these green business practices are illustrated in Figure 12.3 [12]. Your business will have a greater chance of long-term success if you follow the following eco-friendly practices:

- Reduce your consumption of natural resources.
- Reduce energy consumption and increase energy efficiency.

FIGURE 12.3
Green business practices [12].

- Turning off electronic devices when not in use.
- Reduce waste and decrease costs.
- Reduce costs with efficient technology.
- Recycling reduces your costs.
- Good practice of green business can attract new customers.
- Earn a green reputation for your business.
- Work from home if possible.

Examples of leading companies that are going green include Walmart, Apple, Dupont, GE, McDonalds, Home Depot, Wal-Mart, Coca-Cola, Sears, Bank of America, Toyota, Ford, Dell, and Target. Google, eBay, and Facebook are using clean energy sources to run their energy-intensive data centers. Organizations green their businesses by responding to the imperatives from government and stakeholders to implement green practices. They continue to raise the bar and are committed to remove a billion tons of greenhouse gases from the atmosphere. Governments and other policy makers (such as the UN) should proactively encourage corporations to protect the environment.

12.6 Benefits

Some regard green business as smart business. It provides customers with ecological sound products and services. It helps business succeed in that it helps us identify risks and opportunities that would be unnoticed otherwise. The sustainability imperative has the potential to lower the impact of businesses on the environment. Sustainability is good for business because it offers a vitally important business goal for stakeholders, including investors, customers, and policy makers. It also ensures that future generations may not experience resources scarcities. To achieve sustainability, organizations must consider social and environmental issues in their business planning [13]. Being aware of this, more companies are bringing green products to the market and are implementing eco-friendly activities. In this age of sustainability, green business models potentially perform much better compared to the conventional business models [14]. Green business inherently contributes to green jobs, decent work, and the development of sustainable solutions to environmental challenges. Green business practices, even if it is just recycling, can have a significant effect on a company's bottom line. Other benefits of greener business include access to untapped markets, improved resource efficiency, and cost savings, and productive and healthy employers [15].

12.7 Challenges

Sometimes green business is difficult to develop because it requires support from a broad range of actors. The environmental impact may influence the costs, quality, or time in positive as well as negative ways. Being able to define green metrics for process, service, and people is crucial to realizing green vision, but it is a challenge to measure characteristics of the "green" business. Many managers face considerable uncertainty over how they can apply green business ideas to enhance their environmental and social practices and sustainability goals.

Many small and medium enterprises seem to lag behind larger ones in the adoption of green business practices. They are less aware of the necessity of green business and are not ready to adopt it [16]. Other barriers include government constraints, financial constraints, company constraints, and changing market dynamics. Regulations influence and distort green business models. Some companies are too conservative to transform their business model since that may involve large costs of new machinery and new materials. Employees may lack knowledge and skills in the development and production phases. Some customers may not know enough about sustainability and refuse to change their buying habits. The benefits of going green for a business far outweigh the negative issues.

12.8 Conclusion

The environmental impact of doing business becomes increasingly a concern for organizations and their customers. The message to business leaders is clear and simple: Take stock of your company's climate as well as its social impacts. Green business is of global concern. Individuals, communities, and corporations must change patterns of behavior and interactions to create a sustainable future. More and more organizations are cautious about their environmental performance. They not only seek to comply with environmental regulations, but they discover that being green has impact on corporate profitability as well. They demonstrate their environmental awareness and commitment through green buildings, eco-labels, industry pledges, and clean technologies.

Green businesses create jobs that empower workers and honor their humanity. They ensure that they do more with less and use the safest ingredients to keep their customers healthy. They also provide green living alternatives to improve quality of life and avoid polluting their local communities. They play a key role in the transformation of our society to one that is socially just and environmentally sustainable.

Universities around the world are beginning to acknowledge green business education as part of their curriculum [17]. To learn more information about green business, one should consult books in References [5, 11, 18–22].

References

1. H. Yi, "Green business in a clean energy economy: Analyzing drivers of green business growth in U.S. states," *Energy*, vol. 68, 2014, pp. 922–929.
2. "E-commerce business or traditional business: A comparison," https://www.shopaccino.com/blog/ecommerce-business-or-traditional-business-a-comparison
3. S. Kabiraj, V. Topkar, and R. C. Walke, "Going green: A holistic approach to transform business," *International Journal of Managing Information Technology*, vol. 2, no. 3, August 2010, pp. 22–31.
4. A. K. Townsend, "An assessment and critique of green business best practices," *Doctoral Dissertation*, Antioch New England Graduate School, April, 2004.
5. S. Seidel, J. Recker, and J. vom Brocke (eds.), *Green Business Process Management*. Berlin: Springer, 2012.
6. "Sustainable business," *Wikipedia*, the free encyclopedia https://en.wikipedia.org/wiki/Sustainable_business
7. R. Mintzer, *Start Your Own Green Business*. Irvine, CA: Entrepreneur Press, 2009.
8. P. Lago and T. Jansen, "Creating environmental awareness in service oriented software engineering," in E. M. Maximilien et al. (eds.), *Service-Oriented Computing*. Berlin, Gremany: Springer-Verlag, 2011, pp. 181–186.
9. L. Viswanathan and G. Varghese, "Greening of business: A step towards sustainability," *Journal of Public Affairs*, vol. 18, 2018, e1705.
10. W. Ashton, S. Russell, and E. Futch, "The adoption of green business practices among small US Midwestern manufacturing enterprises," *Journal of Environmental Planning and Management*, vol. 60, no. 12, 2017, pp. 2133–2149.
11. G. Weybrecht, *The Sustainable MBA: A Business Guide to Sustainability*, 2nd ed. Chichester, UK: John Wiley & Sons, 2014, p. 117.
12. L. Cekanavicius, R. Bazyte, and A. Dicomanaite, "Green business: Challenges and practices," *Ekonomika*, vol. 93, no. 1, 2014, pp. 74–812.
13. S. Nair and H. Paulose, "Emergence of green business models: The case of algae biofuel for aviation," *Energy Policy*, vol. 65, 2014, pp. 175–184.
14. J. H. Chen and S. I. Wu, "A comparison of green business relationship models between industry types," *Total Quality Management & Business Excellence*, vol. 26, no. 7–8, 2015, pp. 778–792.
15. International Labour Office, *Green Business Booklet*. Geneva: ILO, 2017.
16. S. H. Chuna, H. J. Hwang, and Y. H. Byun, "Supply chain process and green business activities: Application to small and medium enterprises," *Procedia—Social and Behavioral Sciences*, vol. 186, 2015, pp. 862–867.

17. M. Gonglewski and A. Helm, "LaissezFair: A case for greening the business German curriculum," *Die Unterrichtspraxis/Teaching German*, vol. 46, Fall 2013, pp. 200–214.
18. N. Gunningham, R. A. Kagan, and D. Thornton, *Shades of Green: Business, Regulation, and Environment*. Stanford, CA: Stanford University Press, 2003.
19. A. Sommer, *Managing Green Business Model Transformation*. Berlin: Springer, 2012.
20. D. Koechlin and K. Muller (eds.), *Green Business Opportunities*. London: Pitman Publishing, 1992.
21. J. V. Brocke, S. Seidel, and J. Recker, *Green Business Process Management: Towards the Sustainable Enterprise*. Berlin: Springer-Verlag, 2012.
22. P. O. de Pablos, *Green Technologies and Business Practices: An IT Approach*. Hershey, PA: IGI Global, 2011.

13

Green Marketing and Greenwashing

> Success is not the key to happiness. Happiness is the key to success.
> If you love what you are doing, you will be successful.
>
> — **Albert Schweitzer**

13.1 Introduction

The growing international concerns about the environmental sustainability (environment, economy, and social justice) and the future of our planet have compelled companies worldwide to incorporate environmental issues in their business strategies. Governments worldwide are making efforts to minimize human impact on the environment. Consumers are also aware of the environmental issues like global warming and environmental pollution. Environmentalism has emerged to be a vital business movement and is impacting business choices.

Green marketing (or environmental marketing, ecological marketing) involves integrating the concerns about the environment into the practice and principles of marketing. It refers to the process of selling products/ services based on their environmental benefits. It is done by the marketers to introduce their green products to the consumers. Green marketing practices include product design innovations, responsible sourcing, recycling practices, ethical standards, responsible advertising, green communication practices, sustainable packaging, and production of green products. Many companies are moving toward green marketing as part of their organization's overall social responsibility.

The growing demand for greener products and services has led to greenwashing. Greenwashing is making exaggerated or misleading environmental claims in order to curry consumer favor. It is overstatement of environmental attributes in marketing products. It is a breach of marketing ethics. More and more companies are engaging in greenwashing about the environmental benefits of their products or services. They use unique techniques to lure in customers.

This chapter introduces two concepts: green marketing and greenwashing. It begins with the concept of green marketing and its examples and addresses the advantages, disadvantages, benefits, and challenges of green marketing. This chapter also introduces the greenwashing phenomenon and its seven sins, examples, and consequences and highlights how to combat greenwashing and its challenges. The last section concludes the chapter.

13.2 Conventional Marketing

Marketing is a function that enables a company to create, communicate, and deliver products or services to its customers. It is a discipline where the traditional concept states that goods are produced according to the needs and demands of the customers. Marketers assert their influence throughout the lifecycle of a product and use their power to inspire positive changes in the consumer behavior. Capitalism dictates increasing pressure for profits at any cost. Marketing has been transformed during the past few decades through the introduction of modern Information Communications and Technology (ICT), changing consumer behavior and global supply chains.

Modern marketing has created a lot of problems. Giant factories, in particular, have become the source of excessive pollution. Marketers have often been blamed in a general way for the resulting environmental damage. As someone well said, "Marketers are blamed for a multitude of sins: encouraging ever greater consumption of alcohol, fatty foods, empty calories, water and biological resources; using too much packaging; limiting the useful life of products so that people are forced to replace them earlier than necessary; producing greenhouse gases. The list seems never-ending" [1].

A large number of businesses have employed strategies to address the growing concern for human health and the natural environment. Customers too are becoming more demanding and they have started to pay more attention to the environment. Today, concepts such as green marketing, green business, and green products appear more frequently. As society becomes more concerned with the natural environment, businesses have begun to modify their behaviors in an attempt to address society's new concerns. Business as usual is not compatible with a sustainable future.

13.3 Green Marketing Concept

The concept of green marketing has evolved from the traditional marketing since the late 1980s. The differences between traditional and green marketing are illustrated in Figure 13.1 [2]. Green marketing is a marketing philosophy

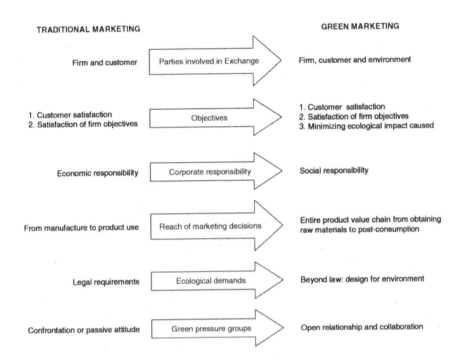

FIGURE 13.1
Differences between traditional and green marketing [2].

that promotes production, consumption, and disposal of eco-friendly products. Green indicates pure in quality and fair in dealing. It refers to marketing products or services based on their environmental benefits. It is mainly concerned about the environment and follows the concept of social marketing. It is practiced by companies that embrace sustainable business practices and corporate social responsibility.

The green marketing mix comprises five dimensions, namely, 5Ps [3], which are explained below and illustrated in Figure 13.2 [4]. Green marketers must address the 5Ps in innovative ways.

- *Product:* Green marketing should begin with green product and service. A green product is one that provides environmental, social, and economic benefits over its lifecycle. A producer should offer ecological products. Companies are now offering more eco-friendly alternatives to their customers. Marketing green products is different from making non-green products. Some common characteristics of products generally accepted as green are [5]: (1) energy efficient, (2) water efficient, (3) low emitting, (4) safe and/or healthy products, (5) recyclable and/or with recycled content, (6) durable, (7) biodegradable, (8) renewable, (9) reused products, (10) third party certified to public or transport standard, and (11) locally produced.

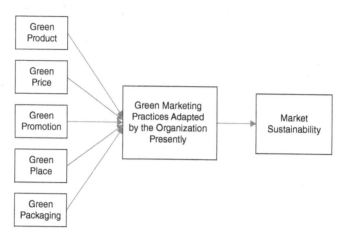

FIGURE 13.2
Green marketing concept [4].

- *Price:* Prices for such products may be a little higher than conventional alternatives. Products are marketed through competition based on price. Customers who are more receptive to environmentally friendly products are often more willing to pay extra for them.

- *Place:* The decision on where and when to make a product available has a significant impact on the customers to attract. Marketing local and seasonal products, for example, vegetables from regional farms, is easier to be marketed "green" than the imported products.

- *Promotion:* Publicizing the green characteristics of the product. Ecological products will probably require special sales promotions. Green marketing is the promotion of a product based on its environmental performance. Promoting products and services to target markets includes advertising, public relations, sales promotions, and direct marketing.

- *Packaging:* Packaging gives the first impression of the product that consumers have. Green packaging is also known as ecological packaging. Sustainable packaging is a development and improved eco-friendly packaging products are used for this type of packaging. Sustainable packaging uses environmentally sensitive methods.

Environmental labels (or eco-labels) are labels used to help the customer distinguish the environmentally friendly products or services from conventional ones. Eco-labels (or sustainable labels) are considered to be an effective tool to market greener products and services that are eco-friendly.

FIGURE 13.3
Typical eco-labels [6].

Without eco-labeling standards, consumers will not be able to tell which products and services are truly beneficial. Some typical eco-labels are shown in Figure 13.3 [6].

The American Marketing Association (AMA) defines green marketing in three ways [7]:

1. *From the retailing aspect:* The marketing of products that are presumed to be environmentally safe.
2. *From social marketing aspect:* The development and marketing of products designed to minimize negative effects on the environment or to improve its quality.
3. *From the environments aspect:* The efforts by organizations to produce, promote, package, and reclaim products in a manner that is sensitive to ecological concerns.

Although the idea of sustainable development appeared in the 1970s, only recently it has been incorporated by businesses. The term "green marketing" was first introduced by Lazez in 1969 as a social dimension of marketing strategy. In the 21st century, corporations are under pressure to address environment issues. Green marketing has become important strategy for companies to remain profitable and competitive. It must meet two objectives: improved environmental quality and customer satisfaction. Several big brands have started to design and manufacture greener products.

13.4 Examples of Green Marketing

Green marketing appeals to the desires of environmentally concerned customers. It has been successfully applied in automobile industry, hotel industry, chemical industry, real estate enterprise, manufacturing, information
and communication, pharmaceutical, and medical services.

- *Hotel industry:* Many hotels have responded to environmental problems by implementing environmental programs. Hotels promote
  their green products and services, such guestrooms, the organic food
  ingredients, and beverage outlets. Many hotels strive to increase
  their bottom lines with different environmental services designed
  to improve the green image of their hotels [8].

- *Automobile industry:* Given the increasing environmental awareness,
  automakers are poised to introduce innovative green gasoline-based
  cars. For example, Volkswagen desires to use hydrogen as a fuel [9].
  Excessive automobile consumption is not an ecologically sustainable
  option for any nation.

- *Real estate:* Real estate enterprise has taken green marketing as a part
  of their strategy to promote their products by employing environmental claims. Green real estate is the future development trend of
  the real estate. Green marketing brings new market opportunities to
  real estate enterprise [10].

Many companies, such as automobile, food, real estate, consumer electronics,
and housing, have begun to adopt green marketing practices and are behaving in an environmentally responsible fashion. These include Dell, McDonald,
Coca-Cola, Wal-Mart, Toyota, 3M, Canon, Xerox, HP, Philips, and Bank of
America. For example, Body Shop offers consumers environmentally responsible alternatives to conventional cosmetic products. Coca-Cola has invested
heavily in various recycling activities and has modified its packaging to minimize its environmental impact. Xerox introduced a recycled photocopier paper
in place of firms which are less environmentally harmful products [11].

13.5 Advantages and Disadvantages of Green Marketing

The marketing strategies for green marketing include the following points [12]:

1. It ensures sustained long-term growth along with profitability.
2. It saves money on advertisement in the long run, although initial
   cost is more.

3. It helps the companies to market their products and services keeping the environment aspects in mind.

4. It also makes most of the employees feel proud and responsible to work for an environmentally responsible company.

5. It promotes corporate social responsibility and improve corporate image.

Disadvantages of green marketing include [13]:

1. Changes leading to cost
2. Costly green certificates
3. Availability of identical products in the market
4. Greenwashing

13.6 Benefits and Challenges of Green Marketing

Green marketing is marketing products that are considered to be environment-friendly. The main objective of any green marketing measure is to reduce the organization's environmental impact. Green marketing can also provide important benefits [14].

- Cost reduction resulting from lower water, energy, and other resources consumption.
- Resource saving due to material recycling.
- Profit gain due to residual reuse.
- Detection of new raw materials and manufacturing processes.
- "Clean" manufacturing technology patents' sales.
- Firm image improvement and sales increase, due to of ecological products development and launch.
- Possibility of entering in the international market, increasingly rigid in regards to environmental restrictions.
- Greater facility of obtaining foreign financing.
- Greater acceptability of shareholders who prefer to invest in environmentally responsible firms.

As green marketing becomes an essential tool for sustainable business strategy, companies are adopting green marketing practices. However, green marketing has not achieved its potential for improving the quality of life for consumers [15]. Challenges of green marketing include [13, 16]:

1. Green products require renewable and recyclable material, which is costly.
2. There are problems of deceptive advertising and false claims.
3. It requires a technology, which requires huge investments in research and development.
4. Majority of the people are not aware of green products and their uses.
5. Majority of the consumers are not willing to pay a premium for green products.
6. It is challenging to educating customers about the advantages of green marketing.
7. There is a lack of standardization to authenticate claims or certify a product as green.
8. There is huge investment in R&D.
9. Green products are generally perceived to be more expensive than conventional choices and consumers are reluctant to pay higher price.
10. There is an effort to replace conventional products and a lack of confidence.
11. There are problems of deceptive advertising and false claims (or greenwashing).

Many consumers will choose products that do not damage the environment over less environmentally friendly products, even if they cost more. For some consumers, the environmental benefit outweighs the price difference.

Some consumers are skeptical and confused about green product claims and the truth of the green messages they receive. The act of giving lip service to loving the environment or presenting a product as green when it is not called greenwashing. In spite of these challenges, green marketing has continued to gain adherents.

13.7 Greenwashing Concept

The increasing popularity of green products and services has led to greenwashing. Greenwashing is the practice of making unwarranted or overblown claims of environmental friendliness in an attempt to sell a product or service. It refers to advertising that cheats consumers about a product's environmental features. The ads and labels that promise more environmental benefit than they deliver is greenwashing. Greenwashing also occurs when companies deliberately frame their activities as "green" in order to look environmentally friendly. As the eco-friendly/natural/green movement grows, the more marketers are going to use it to their advantage.

The term "greenwashing" was coined in 1986 by Jay Westerveld, a field biologist and activist. The act of greenwashing, also known as "green sheen, green PR, or green marketing," involves misleading consumers about the environmental benefits of a product through false or misleading advertising. It may be misleading consumers and investors by telling the truth, but not the whole truth. There are four types of greenwashing being under attack: the greenwashing of products, processes, symbols, and structures. Green marketing is not simply giving lip service to loving the environment, not backed with action. Greenwashing can damage a business reputation; it is a risky course of action.

Conventional wisdom suggests that greenwashing should be regulated because greenwashing misleads consumers to choose non-green products. To that end, the US Federal Trade Commission has prohibited greenwashing [17]. Greenwashing hinders green marketing since it makes consumers to be skeptical about their green claims. Although greenwashing is not a new phenomenon; its use has increased in recent years to meet consumer demand for environmentally friendly products and services. Greenwashing merges the concepts of "green" (environmentally sound) and "whitewashing" (to gloss over wrongdoing). It started with the rise of the environmental movement in the mid-1960s.

13.8 The Seven Sins of Greenwashing

Greenwashing takes place when a company spends more time and money claiming to be environmentally responsible and "green" rather than actually implementing this into their business practices. TerraChoice identified what it calls the Six Sins of Greenwashing, which are stated as follows [18, 19]:

1. *Sin of the hidden trade-off:* This is committed by suggesting a product is "green" based on an unreasonably narrow set of attributes without attention to other important environmental issues.

2. *Sin of no proof:* This is committed by an environmental claim that cannot be substantiated by easily accessible supporting information.

3. *Sin of vagueness:* This is committed by every claim that is so poorly defined that its real meaning is likely to be misunderstood by the consumer.

4. *Sin of irrelevance:* This is committed by making an environmental claim that may be truthful but is unhelpful for consumers seeking environmentally preferable products.

5. *Sin of lesser of two evils:* This is committed by claims that may be true within the product category, but that risk distracting the consumer from the greater health.

6. *Sin of fibbing:* This is committed by making environmental claims that are simply false. This is the least frequent sin.

7. *Sin of false labels:* This is committed by exploiting consumers' demand for third-party certification with fake labels.

13.9 Examples of Greenwashing

"Eco-friendly," "organic," "natural," "sustainable," "recyclable," and "green" are just some examples of the widely used labels that can be confusing and misleading to consumers. It seems everybody likes the concept of green and uses the term as a sales niche. But what is marketed as green is often not what it purports to be. The following examples illustrate greenwashing.

- A trash bag is labeled "recyclable" without qualification.
- Eco-friendly or organic cigarettes and hybrid SUVs create the illusion of being "greener."
- A water company may tend to overrepresent its greenness.
- Polluters make cynical use of culture as a means of greenwashing their public image.
- A house is being marketed as eco-friendly and energy-saving when it doesn't really deserve that description.
- Volkswagen used software to cheat on vehicle emissions tests.
- In 1991, Mobil Corporation agreed to stop advertising its Hefty brand plastic trash bags as biodegradable.
- Volunteer tourism operators are overpositioning and communicating responsibility inconsistently.

The above examples are just a few of the many. Claims that cannot be supported by verified data may destroy or reduce the credibility and legitimacy of the companies.

13.10 Consequences of Greenwashing

The implications of greenwashing are wide spread. So it is important to consider the ethics of greenwashing and its environmental and social consequences. Some companies do not "walk the talk" and do not play by the rules but resort to greenwashing. When the greenwashing behaviors of a company are exposed, the company's business suffers significant losses. The company should provide compensation for stakeholders that will satisfy their expectations. It should also improve their green brand legitimacy to achieve social support [20]. If your company has a reputation for harming the environment, it will take some time before people can trust the company. The best policy is to clean up one's environmental act. Consumers may investigate why certain companies engage in greenwashing and may develop skeptical cognitions and negative attitude toward the companies. Greenwashing has caused consumers to question corporate honesty. Consumers may be confused about which products are actually eco-friendly. Government regulation sets pressures on advertisers to avoid greenwashing. But companies are not required by law to publish their environmental policy statement. Publishing such statement is voluntary.

13.11 Combating Greenwashing

There has been a significant increase in the use of greenwashing in recent years. Allegations of greenwashing have been made against companies, organizations as well as military establishments. If greenwashing practices continue to go unchecked, consumers will become cynical about green claims. Companies, organizations, and individuals are making effort to reduce the impact of greenwashing by exposing it to the public. These days, consumers have become better informed about the choices they make regarding the products they purchase. But the informedness of the consumer may vary from individual to individual.

Although a company's product may seem eco-friendly, it is always best to look behind the "green" facade to get the facts. Companies need to be transparent and provide consumers with the whole truth. They should be aware of the ramifications of deceptive environmental claims. Company perceived as committed to sound environmental policies gains the good graces of consumers. If a company claims to be green, double check their claim. Visit to their website. Is the company overstating their intention? Not all eco-labeling is greenwash. You can visit http://ecolabelling.org/ and search more than 300 labels. The site explains what products the label is used for. It may take some work to figure out which companies are actually being truthful.

13.12 Challenges of Greenwashing

One major result of greenwashing is public confusion. Greenwashing either confuses or misleads people by providing misleading environmental claims. Since people are not well-equipped to verify deceptive advertising and labeling, they end up buying products that don't have the environmental performances that they expect. The skeptical consumers will find it hard to accept advertising claims at face value.

Another challenge is that there is no guarantee that greenwashing companies keep up their good practices. Greenwashing cannot be properly understood without its perception in the eye of the beholder [21].

13.13 Conclusion

The green movement is expanding rapidly worldwide. Green marketing is the marketing of environmentally friendly products and services. It is a tool for protecting the environment for the future generation. It may be regarded as a subset of corporate social responsibility (CSR) strategies. It is all marketing activities geared toward protecting the environment. It is a practice whereby companies seek to go above and beyond traditional marketing by promoting environmental core values. It should be regarded not solely as an activity, but also as a philosophy. It is the path to sustaining success. It is a modern concept for marketers all over the world.

Greenwashing is making unsubstantiated or misleading claims about the actual environmental benefits of a product, service, technology, or company. It appears to be evident in most marketing practices around environmental issues. It only benefits a company when the market is relatively uninformed. It is a phenomenon that persists in corporate marketing and that existing laws are not sufficient to prevent it.

More information on green marketing and greenwashing can be found in the books in References [18, 22–25] and five related journals:

- *Journal of Marketing*
- *Journal of Strategic Marketing*
- *Electronic Green Journal*
- *European Journal of Marketing*
- *International Marketing Review*

References

1. G. Weybrecht, *The Sustainable MBA: A Business Guide to Sustainability*, 2nd ed. Chichester, UK: John Wiley & Sons, 2014, p. 218.

2. A. Chamorro and T. M. Bañegil, "Green marketing philosophy: A study of Spanish firms with ecolabels," *Corporate Social Responsibility and Environmental Management*, vol. 13, 2006, pp. 11–24.

3. "Green marketing," *Wikipedia*, the free encyclopedia https://en.wikipedia.org/wiki/Green_marketing

4. G. Devakumar, et al., "An empirical study on green marketing strategies for market sustainability with respect to organic products," *UAS JMC*, vol 3, no. 2, pp. 33–38. Also available: http://www.msruas.ac.in/pdf_files/Publications/MCJournals/August2017/Paper7.pdf

5. M. Bhatia and A. Jain, "Green marketing: A study of consumer perception and preferences in India," *Electronic Green Journal*, vol. 1, no. 36, 2013.

6. H. M. S. S. Ribeiro, "The importance of green marketing for Portuguese companies in the footwear industry, *Masters Thesis*, ISCTE Business School, September, 2017.

7. W. M. Wong, "To integrate green marketing into software development company's marketing plan as a case studying of FBIC," *Journal of Global Business Issues*, vol. 2, no. 2, Summer 2008, p. 223.

8. E. S. Chan, "Green marketing: Hotel customers' perspective," *Journal of Travel & Tourism Marketing*, vol. 31, no. 8, 2014, pp. 915–936.

9. S. Singh, D. Vrontis, and A. Thrassou, "Green marketing and consumer behavior: The case of gasoline products," *Journal of Transnational Management*, vol. 16, no. 2, 2011, pp. 84–106.

10. "Game analysis of real estate enterprise to implement green marketing," *International Conference on Information Management, Innovation Management and Industrial Engineering*, 2011, pp. 348–351.

11. A. Verma, "Green marketing: Importance and problems associated," *International Journal of Business Management*, vol. 2, no. 1, 2015, pp. 428–437.

12. "5 effective green marketing strategies," https://www.firstcarbonsolutions.com/resources/newsletters/july-2016-effective-green-marketing-strategies/5-effective-green-marketing-strategies/

13. P. Kumar, "Green marketing," *Proceedings of National Conference on Marketing and Sustainable Development*, October 2017.

14. L. Simão and A. Lisboa, "Green marketing and green brand—The Toyota case," *Procedia Manufacturing*, vol. 12, 2017, pp. 183–194.

15. K. Papadasa, G. J. Avlonitisb, and M. Carriga, "Green marketing orientation: Conceptualization, scale development and validation," *Journal of Business Research*, vol. 80, November 2017, pp. 236–246.

16. S. Chand, "Green marketing: Evolution, reasons, advantages and challenges," http://www.yourarticlelibrary.com/marketing/green-marketing-evolution-reasons-advantages-and-challenges/32326

17. H. C. B. Lee, J. M. Cruz, and R. Shanka, "Corporate social responsibility (CSR) issues in supply chain competition: Should greenwashing be regulated?" *Decision Sciences*, vol. 49, no. 6, 2018, pp. 1088–1115.

18. TerraChoice, *The Sins of Greenwashing: Home and Family Edition*. Northbrook, IL: Underwriters Laboratories, 2010.

19. R. Dahl, "Green washing: Do you know what you're buying," *Environmental Health Perspectives*, vol. 118, no. 6, June 2010, pp. A246–A252.

20. "Timely or considered? Brand trust repair strategies and mechanism after greenwashing in China—From a legitimacy perspective," *Industrial Marketing Management*, vol. 72, 2018, pp. 127–137.

21. P. Seele and L. Gatt, "Greenwashing revisited: In search of a typology and accusation-based definition incorporating legitimacy strategies," *Business Strategy and the Environment*, vol. 26, 2017, pp. 239–252.

22. M. J. Baker (ed.), *The Marketing Book*, 5th ed. Oxford: Butterworth-Heinemann, 2003.

23. J. Ottoman and E. S. Miller, *Green Marketing Opportunities for Innovation*. New York, NY: McGraw-Hill, 1991.

24. F. Bowen, *After Greenwashing: Symbolic Corporate Environmentalism and Society*. Cambridge, England: Cambridge University Press, 2014.

25. T. Panayotou, *Green Markets: The Economics of Sustainable Development*. San Francisco, CA: Press Institute for Contemporary Studies, 1993.

14

Green Manufacturing

Great minds have purposes, others have wishes. Little minds are tamed and subdued by misfortune, but great minds rise above them.

—Washington Irving

14.1 Introduction

The rapid technological advancements have led to a growing concern for environmental degradation caused by the manufacturing sector. Manufacturing is well known as the largest sector of the American economy. It is closely connected with all other sectors like mining, trading, supply chain, and financial services. It contributes to the economy by providing many job opportunities, creating wealth, eradicating poverty, providing better life standards, healthcare, and education [1]. However, the manufacturing sector accounts for a significant portion of the world's consumption of resources and generation of waste. It has negatively impacted the environment through the overexploitation of natural resources and pollution. The manufacturing industry's energy demand is one-third of the total energy consumption in the United States [2]. To minimize the environmental damage due to manufacturing requires a new manufacturing process.

Green manufacturing (GM), also known as environmentally conscious manufacturing, is the embodiment of the strategy for sustainable development of the manufacturing sector. Green manufacturing refers to modern manufacturing that makes products without pollution. It alleviates the current contradiction between industrial development and environmental pollution. It addresses a wide range of environmental and sustainability issues, including resource selection, transportation, manufacturing process, and pollution. This new way of thinking about manufacturing is having a big impact on manufacturers worldwide as they realize the many financial benefits of adhering to sustainable principles.

Green manufacturing has attracted the attention of industries all over the world and consumers are demanding it. Its main objective is to reduce environmental waste. It can be applied in all manufacturing sectors that minimize waste and pollution and conserve resources. Green manufacturing is an effective way to protect resources and environment and conserve the resources for future generation.

This chapter provides a brief introduction to green manufacturing, an area of great importance for current and future manufacturing operations. It begins with traditional manufacturing concepts and then addresses green manufacturing practices. This chapter also highlights some applications of green manufacturing and discusses how to improve awareness about green manufacturing and covers the benefits and challenges of green manufacturing. The last section concludes the chapter.

14.2 Traditional Manufacturing

Manufacturing is the way of transforming resources into products or goods, which are required to cater to the needs of the society. It started from a small-scale production line of crafts in the 1800s. It has evolved into large-scale mass production. The 2000s is the era when computerized and personalized manufacturing systems came into existence for mass production lines [3]. As illustrated in Figure 14.1 [4], manufacturing is typically a series of processes comprising of selection of raw materials, production of objects, assembling

FIGURE 14.1
Stages in the manufacturing process [4].

of parts, inspection, and dispatching. It may include foundry, forging, jointing, heat treatment, painting, etc. It involves using resources to meet the delivery date, cost, quality, and optimal economic goals in limited resource conditions. It often involves mass production and heavy energy consumption such as coal or electricity.

The traditional manufacturing processing techniques consume lot of energy, mainly for production and utility. They also produce a lot of pollution and add to the deterioration of the global environment. Because of this, manufacturers are gradually transforming their manufacturing systems from traditional mass production to flexible lean systems. With rapid changes in technology, manufacturing itself is constantly transforming and evolving, as illustrated in Figure 14.2 [5]. It now takes a proactive role in the development of cleaner manufacturing processes. In order to minimize the environmental damage due to manufacturing, there is a need of new manufacturing process.

In modern times, two types of manufacturing systems have emerged, emphasizing waste minimization. They are "lean" manufacturing systems and "green" manufacturing systems and both reduce waste. Lean manufacturing seeks to eliminate all types of wastes generated within a production system. In lean manufacturing, there are eight categories of waste that should be monitored [6]: (1) overproduction, (2) waiting, (3) inventory, (4) transportation, (5) over-processing, (6) motion, (7) defects, and (8) workforce. Lean and green best practices are considered complementary, as shown in Figure 14.3 [7]. Lean manufacturing is transcending to a more green state. Green manufacturing is an advanced, modern manufacturing approach that comprehensively considers the environmental influence and the resources consumption. It differs from traditional manufacturing in that it focuses on environmental impact.

14.3 Green Manufacturing Practices

The term "green manufacturing" originated in Germany in the late 1980s. Green manufacturing is involved in the whole product lifecycle. It is a step toward sustainable manufacturing. As illustrated in Figure 14.4 [8], green manufacturing has to address all five areas: green resources, green design, green production, green manufacturing, and green disposal.

- *Green resources:* Green manufacturing uses energy, water, and other resources more efficiently thereby reducing the overall impact on the environment. Workers use fewer natural resources. Using fewer resources to make the same product saves money. Workers may generate electricity from renewable sources which include wind, biomass, geothermal, solar, ocean, and hydropower.

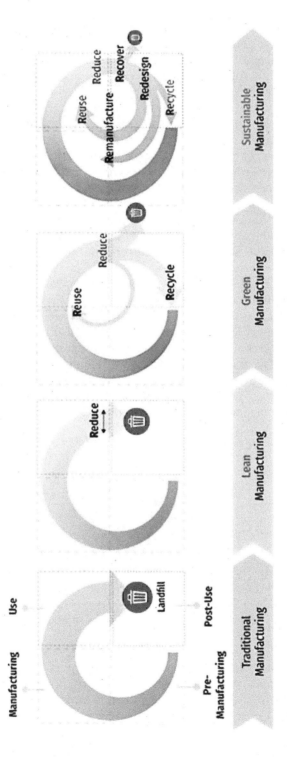

FIGURE 14.2
Evolution of manufacturing [5].

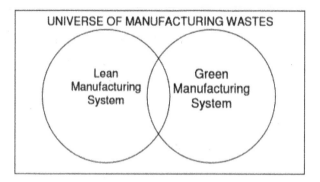

FIGURE 14.3
Lean and green best practices are complementary [7].

- *Green design:* This is also known as environment conscious design or product lifecycle design. Green design focuses on developing eco-friendly products that minimize waste. It considers the environment factors and measures pollution prevention in the product design phase. Product designer needs to consider the manufacturability of the product, the energy consumption, the maintainability, and reusability of the product [9]. If environmental protection is considered at the design stage of a product, the pollution can be easily reduced

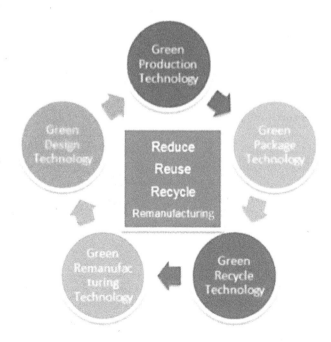

FIGURE 14.4
Principles of green manufacturing [8].

to the minimum. Manufacturing companies need to shift toward using cleaner energy.

- *Green production:* Green technologies focus on reducing the impact of manufacturing processes at every stage of production. Companies should stay ahead of the curve on sustainability and evaluate how green are the products, resources, and energy. The selection of raw materials directly dictates the realization of the green production. A growing range of green products, ranging from organic food products to electric cars, are being offered by companies to customers [10]. By developing green products that are demanded by consumers, companies can derive additional sales that can offset their cost of development. Green products save energy, reduce energy consumption, and increase the company's competitive advantage.

- *Green manufacturing:* Green manufacturing considers the impact of the product development, manufacturing, and activity on the environment. This ensures sustainability in resource extraction, material processing, product use, and disposal. The green manufacturing process obeys the following five principles [11]: (1) least resources consumption principle, (2) least energy consumption principle, (3) least environment pollution principle, (4) better labor protection principle, (5) economic efficiency principle.

- *Green disposal:* Green disposal aims to reduce e-waste by repairing, redeploying, disposing, refurbishing, retaining, and reusing of outdated IT hardware. Reducing resource use, waste, and pollution, along with recycling and reusing waste, yields benefits. Waste is whatever does not add value to the end product. To achieve green disposal, release of toxic substances in product life is not allowed. Overproduction, which is a form of waste, occurs when production output exceeds actual customer orders.

The conventional manufacturing wisdom suggests that we cannot improve what cannot be measured. There is the need for developing metrics that track and monitor performance of green manufacturing. A toolbox (Greenometer) to assess the greenness level of manufacturing companies has been proposed. Greenometer offers relative greenness assessment among different industries [12].

14.4 Green Manufacturing Strategies

Green manufacturing is somewhat a philosophy for manufacturing that minimizes waste and pollution through product and process design. The main objective of green manufacturing is sustainability. Today, manufacturers must keep the environment in mind at each step of their product lifecycle. To go green, one should consider the following manufacturing strategies [13]:

1. Design for disassembly, remanufacture, or reuse
2. Rethink product and process technology
3. Streamline the supply chain
4. Reduce energy and water consumption
5. Choose recyclable or biodegradable materials and packaging.
6. Integrate environmental costs into your production budget
7. Use ISO 14001 standard as a jumping-off point.
8. Find a reverse logistics supply vendor.
9. Invest in business intelligence/analytics
10. Redesign all scales of manufacturing flow
11. Shift to a service-oriented business

Some organizations have started developing competitive green manufacturing strategies. This enhances the image of companies in the eyes of customer and their competitiveness. Green manufacturing strategic patterns adopted by ISO 14001 certificate holders in Jordan were agile, lean, and care-taker strategic patterns. The green strategic patterns are the milestones of the green practices success. The effectiveness of adopted green strategies are yet to be determined [14].

14.5 Motivations for Green Manufacturing

The manufacturing industry is motivated to adopt green manufacturing practices to reduce the environmental impact and improve the economic performance. Such practices include pollution prevention, product stewardship, and emission control. There are a number of factors that motivate the manufacturing industry to implement green manufacturing. These factors can be grouped into the three categories of regulatory pressure, economic incentives, and competitive advantages. These factors include [15]:

1. Pressure from government—regulations, and tax benefits
2. Access to government incentives
3. Increase sales
4. Save money on energy costs
5. Boost employers' morale
6. Interest in efficiency

7. Scarcity of resources
8. Pressure from society/consumers and competitors
9. Desire to maintain market leadership
10. Ensure control of supply chain effects

US governmental agencies have developed a series of policies, regulations, and laws, which has achieved significant progress in advancing the environmental performance of manufacturers. These governmental efforts compel the manufacturing industry to consider green manufacturing as the economic benefits, which could result from the implementation of sustainability program. For example, the US Food and Drug Administration (FDA) is responsible for evaluating the safety and efficacy of new drugs.

14.6 Applications of Green Manufacturing

Green manufacturing refers to the new manufacturing paradigm that employs various green strategies and techniques to become more eco-efficient. This new green manufacturing paradigm is an outcome of market and technological drivers. It may also be regarded as an important component of green business. The drivers that play a crucial role in the implementation of greening throughout manufacturing sectors include financial benefits, company image, environmental conservation, compliance with regulations, stakeholders, green innovation, supply chain requirements, customers, employee demands, internal motivations, market trends, and competitors [16].

Interest in green manufacturing is increasing more and more within the industrial communities [17]. The following are typical applications of green manufacturing [18]. Transportation consumes much of the earth's resources. It is key to greening global industries. A basic change in vehicle design is necessary. It will reduce emissions, fuel consumption, and cost. Additive manufacturing or 3D printer needs to be considered as a practical green process because of saving in materials and reduction in processing steps. The production of semiconductor/electronic products (such as instruments, radio, TV, radars, computers, mobile phones) takes up a great deal of resources as well as generates harmful wastes during their production. The pharmaceutical industry fabricates chemical products and receives strict regulatory oversight to protect the public from unsafe pharmaceutical products. For example, paint manufacturing industry generates large quantities of hazardous and nonhazardous wastes. Waste reduction should be a high priority for this manufacturing sector.

Green manufacturing technique can be used for green industrial facilities by practicing it from workplace to improve the environmental outcomes of production processes. Green manufacturing can be practiced by reducing the cost of raw material, that is, by using less energy and reusing the recycled wastes instead of buying new materials for production [2].

The automotive industry has become indispensable, and it plays a vital role in national economy. The industry is based on the great consumption of resources and energy, while continuous increment of disposed of vehicles leads to severe pollution [8]. Other applications include renewable energy systems, green chemistry avoidance of toxins, furniture industry, textile industry, carpet industry, and cement industry.

Stakeholders including regulators, customers, shareholders, board members, and employees are increasingly demanding companies to be more environmentally responsible with respect to their products. Their reasons include regulatory requirements, product stewardship, enhanced public image, potential to expand customer base, and potential competitive advantages [19].

A lot of companies all over the world claim to be "going green." Examples of companies who are making strides in sustainable/green manufacturing include Coca-Cola, BMW, GM, Ford, Motorola, Toyota, IBM, Dell, Siemens, Samsung, Apple, Nike, Johnson & Johnson, and Tupperware.

14.7 Awareness of Green Manufacturing

The awareness of green manufacturing is crucial to its adoption. In order to adopt and implement green manufacturing, it is important for the society to be aware of its significance. There is no doubt that green manufacturing can be easily accepted and supported by the populace, the government, and the nonprofit organizations. Awareness about green manufacturing must be improved through education. Everyone associated with the business must be involved in adopting green manufacturing, including suppliers, customers, and employees. Training and education programs are thus essential for employees. Students and workers in manufacturing must learn about green manufacturing, sustainability concepts, and practices such as [20]:

1. *Energy from renewable sources:* Workers may generate electricity from renewable sources such as wind, biomass, geothermal, solar, ocean, hydropower, etc.

2. *Energy efficiency:* Workers will utilize specific technologies and practices to improve energy efficiency. More energy-efficient technology will help in using renewable energy and protect the world from climate change.

3. *Pollution reduction:* Workers will use green technologies and practices to reduce the generation of pollutants. Manufacturers must take all reasonable steps to eliminate pollution.

4. *Natural resources conservation:* Workers will use specific technologies and practices to conserve natural resources such as land, water, etc. With resources becoming scarcer due to depletions and exploitations, we must reduce waste by minimizing natural resource use and recycling.

Organizations can assess the commitment of the top management to green manufacturing in terms of availability of an explicit environment policy, availability of strategy to achieve the policy, and allocating financial resources [21].

14.8 Benefits and Challenges

Manufacturing is an important sector due to its contribution to the economy and creation of jobs. Manufacturing companies will adopt green manufacturing if they realize that it will result in some financial benefits. Green manufacturing can mitigate air, water, and land pollution. It reduces waste at the source and minimizes health risks to humans. Adoption of green manufacturing is easily justified considering the benefits. The benefits for engaging in green initiatives include [22]:

- Get tax incentives
- Adhere to compliance regulations
- Save cost
- Reduce impact on the environment
- Improve corporate image
- Maintain competitive advantage
- Increase employee morale
- Attract new customers

Other benefits include lower raw material costs, production efficiency gains, tax incentives, complying with the regulations, and reducing impact on the environment.

In spite of the solution that green manufacturing offers to the environmental problem, it is facing some challenges including:

- Being still skeptical about the business benefits
- The costs of embarking on green manufacturing
- Lacking required skills and shortage of financial resources

- Lacking capable scientifically based decision support tools for effective implementation of green manufacturing
- Often requiring the use of more expensive materials and technologies
- Not keeping pace with the global expansion of the manufacturing industry the rate at which green manufacturing systems are being implemented
- Competing with companies overseas that do not live up to the same standards
- Assessing the performance of green manufacturing

Green manufacturing will have more benefits and challenges in the years to come.

14.9 Conclusion

Green manufacturing has become the inevitable choice in the 21st century manufacturing industry. It is a modern manufacturing approach that gives due consideration to environmental impact and resource consumption. It seems to be the effective way to achieve sustainable manufacturing development. Manufacturing companies worldwide are now conscious about the environment.

Green manufacturing will have a lot of benefits in the years to come. It is the important issue that manufacturing systems of the future must take into account. There are various incentives, resources, and industry-specific organizations available to help manufacturing companies take green initiatives. As long as the benefits of being green outweigh the cost, the adoption of green practices will continue to be attractive and wise. More information about green manufacturing can be found in [15, 23–28] and the journal devoted to it: *International Journal of Precision Engineering and Manufacturing-Green Technology*.

References

1. D. Seth, M. A. A. Rehman, and R. L. Shrivastava, "Green manufacturing drivers and their relationships for small and medium (SME) and large industries," *Journal of Cleaner Production*, vol. 198, 2018, pp. 1381–1405.
2. J. Li et al., "Editorial automation in green manufacturing," *IEEE Transactions on Automation Science and Engineering*, vol. 10, no. 1, January 2013, pp. 1–4.

3. Y. Nukman et al., "A strategic development of green manufacturing index (GMI) topology concerning the environmental impacts," *Procedia Engineering,* vol. 184, 2017, pp. 370–380.

4. D. Swathisri and D. S. S. Kumar, "Green manufacturing technologies—A review," https://www.researchgate.net/publication/305731124_GREEN_ MANUFACTURING_TECHNOLOGIES_-_A_REVIEW

5. "Welcome to the Institute for Sustainable Manufacturing!" https://www.engr. uky.edu/ism

6. G. D. Maruthi and R. Rashmi, "Green manufacturing: It's tools and techniques that can be implemented in manufacturing sectors," *Materials Today: Proceedings,* vol. 2, 2015, pp. 3350–3355.

7. G. G. Bergmiller, "Lean manufacturers transcendence to green manufacturing: Correlating the diffusion of lean and green manufacturing systems," *Doctoral Dissertation,* University of South Florida, October 2006.

8. Z. Zeya, "Green manufacturing framework development and implementation in industry," https://digi.lib.ttu.ee/i/file.php?DLID=3921&t=1

9. W. Qifen, "Green manufacturing-oriented digital system and operation technologies for manufacturing enterprises," *Applied Mechanics and Materials,* vols. 20–23, 2010, pp. 40–44.

10. A. Bhattacharya, R. Jain, and A. Choudhary, "Green manufacturing: Energy, products and processes," March 2011, http://www.cii.in/webcms/Upload/ BCG-CII%20Green%20Mfg%20Report.pdf

11. D. Zhong, "Study on a green manufacturing process design system," *Applied Mechanics and Materials,* vols. 397–400, 2013, pp. 57–61.

12. A. H. Salem and A. M. Deif, "Developing a Greenometer for green manufacturing assessment," *Journal of Cleaner Production,* vol. 154, 2017, pp. 413–423.

13. "Green manufacturing: 8 strategies for success," http://members.questline. com/Article.aspx?articleID=32753&accountID=1&nl=19072

14. Y. K. A. Migdadi and D. S. I. Elzzqaibeh, "The evaluation of green manufacturing strategies adopted by ISO 14001 certificate holders in Jordan," *International Journal of Productivity and Quality Management,* vol. 23, no. 1, 2018, pp. 90–109.

15. D. A. Dornfeld (ed.), *Green manufacturing: Fundamentals and Applications.* New York, NY: Springer, 2013, p. 8.

16. K. Govindan, A. Diabat, and K. M. Shank, "Analyzing the drivers of green manufacturing with fuzzy approach," *Journal of Cleaner Production,* vol. 96, 2015, pp. 182–193.

17. A. M. Deif, "A system model for green manufacturing," *Advances in Production Engineering & Management,* vol. 6, 2011, pp. 27–36.

18. S. H. Ahn, D. M. Chun, and W. S. Chu, "Perspective to green manufacturing and applications," *International Journal of Precision Engineering and Manufacturing,* vol. 14, no. 6, June 2013, pp. 873–874.

19. C. A. Rusinko, "Green manufacturing: An evaluation of environmentally sustainable manufacturing practices and their impact on competitive outcomes," *IEEE Transactions on Engineering Management,* vol. 54, no. 3, August 2007, pp. 445–454.

20. "What is green manufacturing and why is it important?" https://www. goodwin.edu/enews/what-is-green-manufacturing/

21. P. J. Singh and K. S. Sangwan, "Management commitment and employee empowerment in environmentally conscious manufacturing implementation," *Proceedings of the World Congress on Engineering,* London, UK, July 2011.

22. A. Robinson, "Green is the new black: Why green manufacturing & sustainability matter," May 2017, https://cerasis.com/2017/05/17/green-manufacturing/

23. N. K. Jha, *Green Design and Manufacturing for Sustainability*. Boca Raton, FL: CRC Press, 2015.

24. J. Rynn, *Manufacturing Green Prosperity: The Power to Rebuild the American Middle Class*. Santa Barbara, CA: Praeger, 2014.

25. M. J. Franchetti, B. Elahi, and S. Ghose, *Value Creation Through Sustainable Manufacturing*. Maplewood, NJ: Peen Tool Co, 2016.

26. J. K. Wang, *Green Electronics Manufacturing: Creating Environmental Sensible Products*. Boca Raton, FL: CRC Press, 2017.

27. M. Singh, T. Ohji, and R. Asthana, *Green and Sustainable Manufacturing of Advanced Material*. Elsevier, 2015.

28. G. Seliger (ed.), *Sustainable Manufacturing: Shaping Global Value Creation*. Springer, 2012.

15

Green Supply Chain Management

The secret to success is to do common things uncommonly well.

—John D. Rockefeller

15.1 Introduction

Countries all over the world are facing severe problems such as high resource consumption, low efficiencies, and high pollution emissions. There is a growing alarming concern that global warming may lead to a great disaster. People worldwide are expressing concerns over the increase of energy consumption [1]. Developing profitable business without sacrificing the environment is gaining increasing attention.

The supply chain is an important branch of operations management. Businesses no longer compete as autonomous entities as store versus store, but rather as supply chain versus supply chain. The supply chain consists of all parties involved in fulfilling a customer request, including the suppliers, transporters, warehouses, retailers, and customers. Supply chain management (SCM) stands for the chain connecting each element of the manufacturing and supply process from raw materials through to the final customers. It is basically based on interactions of manufacturing, logistics, materials, distribution, and transportation functions within an enterprise.

Traditional supply chains move from the raw material state to the end product. There is a growing need to integrate environmental considerations into SCM. Environmental sustainability practices in the supply chain are often referred to as green supply chain management (GSCM). Sustainable GSCM aims at integrating environmental thinking into SCM. GSCM is an emerging concept motivated by the need for environmental consciousness. It aims at integrating environmental thinking into SCM. Its principles can be applied throughout the entire supply chain including product design, material selection, manufacturing process, and delivery of the final product to consumers.

This chapter provides an introduction to GSCM. It begins with presenting traditional supply chain and addresses green supply chain (GSC). It highlights the motivations for GSC and covers some applications of supply chain. It also discusses SCM and GSCM and elaborates on the benefits and challenges of GSC and GSCM. The last section concludes the chapter.

15.2 Supply Chain

A supply chain is a system that is concerned with transforming materials into a finished product or service. It may also be regarded as a network consisting of all parties involved (e.g., suppliers, manufacturers, transporters, distributors, warehouses, wholesalers, retailers, customers, etc.) in producing and delivery products or services to customers. It is hard to evaluate GSC with single index due to its complexity. Modern supply chain is complex, dynamic, competitive, and flexible. Standard measurements of the performance of the supply chain include customer satisfaction, service, time, responsiveness, cost, and quality.

The concept of supply chain first appeared in the early 1980s. The traditional supply chain integrates raw material delivery, the manufacturing process, and product delivery to customers. Waste is present in the supply chain. Recycling of used products has become an integral component of the supply chain. Flexibility is also an important requirement for supply chain, because it facilitates the response to challenges like globalization and technological changes.

Supply chain reflects a company's capabilities to carry out effective marketing activities. Supply chain practices are used in manufacturing, automotive industry, and healthcare. The conventional supply chain is shown in Figure 15.1 [2]. Geographic information systems (GIS), building information modeling (BIM), and radio-frequency identification (RFID) are currently finding their way into practice in all types of supply chains. The growing awareness of environmental issues has motivated companies to integrate environmentally friendly practices into their traditional supply chain.

15.3 Green Supply Chain Key Concepts

Green supply chain (GSC) is a modern concept that originated in the 1970s. It is regarded as an environmental innovation. GSC is now accepted in many corporate organizations worldwide. It is essentially the extension of supply chain. The GSC includes all components of the conventional supply chain. Adding the "green" component to supply chain involves addressing the relationships between supply chain and the natural environment. GSC integrates environmental parameters (or requirements) into SCM, such as the design of green product, lower costs, and better served customers. The green nature of products is based on the green materials and green technology. Due to the higher cost of making green products, the price of the green products is higher [3].

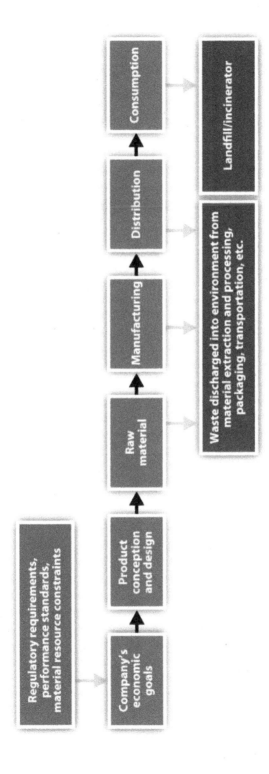

FIGURE 15.1
The conventional supply chain [2].

GSC issues are of prime concern for manufacturing, automotive, electronics, and food industries. A typical GSC is shown in Figure 15.2 [2]. Cooperation among the companies on the supply chain is crucial in improving the environmental compatibility of their businesses. Suppliers, manufacturers, customers, and disposal companies must be integrated in implementing GSC. Sustainable supply chains can only be developed if companies make environmental sustainability a core value not only for their own operations but also for their broader supply chain. Its success highly depends on commitment from the highest levels of a company and also commitment of resources to environmental concerns. If company senior leaders (or top management) fully support sustainability efforts, the rest of the company is more likely to be on the same page. When employees are given freedom to decision making, they work toward enhancing GSC practices.

15.4 Developing Green Supply Chain

Companies must now include "green" or "environmental strategies" in order to retain competitive advantage. The more green a company's supply chain becomes, the more it can become a marketing boon. For green and sustainable product development, it is becoming important to consider social and environment criteria along with economic factors. To make a supply chain green requires that one considers all activities in the supply chain such as raw material procurement, inbound logistics, outbound logistics, marketing, after-sales, and product disposal.

Companies that intend to design and implement a GSC can follow the following basic steps [4].

- *Product selection:* At the product design stage, the product should be designed in such a way that it should be safe for use, creating least pollution and consumes less energy. We always consider development cycle, cost, quality, and other factors and ignore the negative effects like environmental pollution.

- *Process and production:* Process has to be designed so that it conforms to the GSCM initiatives to reduce environmental negative impact. Efficient and effective production strategy to reduce energy consumption that includes reducing waste material.

- *Business partners selection:* This involves electing suppliers or vendors who have proven track records on practicing lean manufacturing and using environment-friendly material.

- *Logistics design:* Efforts should be practiced to reduce fuel consumption. This can be achieved by setting up suppliers near to the OEMs (original equipment manufacturers) and its hubs.

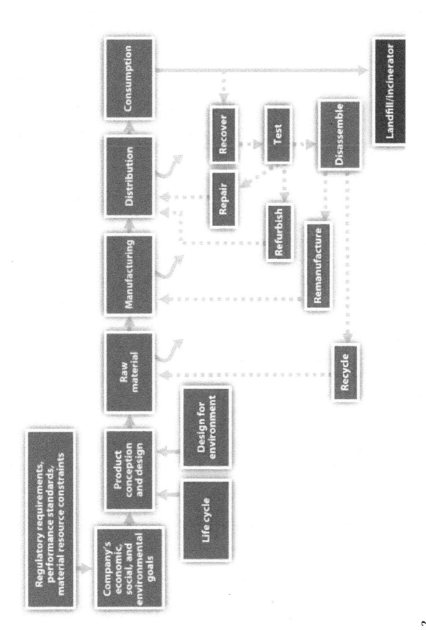

FIGURE 15.2
A typical green supply chain [2].

- *Packaging material:* Replacing package materials that are eco-friendly can be reused and recycled.
- *Information technology:* Green IT (information technology) comprises strategies and best practices for optimizing the usage of computing resources and reducing the environmental footprint of technology.
- *Green building:* Deploying greener practices in design, construction, and maintaining the buildings. Using energy efficient bulbs, natural lightning saves considerable energy. Water has to be recycled for day-to-day use.

Typical examples of GSCs include Walmart, General Electric, Nike, Toyota, British Telecom, and Johnson & Johnson.

15.5 Motivations for GSC

Numerous initiatives have provided incentives for organizations to become more environmentally friendly and adopt GSC. Key internal drivers and external pressures stimulate companies to initiate SCM practices. The key drivers for green initiatives include competition, community pressures, regulation, and government compliance. The risks associated with environmental noncompliance also drive SCM practices. A major motivation for maintaining the sustainability of supply chain is social environmental responsibilities. Another influence is the product life cycle positioning of the products of a company. The product life cycle phase impacts the greening of the supply chain [5].

GSC is becoming popular due to increasing customer awareness. Customers are now demanding green products such as green cars and Energy Star home appliances. Pressure from customers, society, and regulatory agencies is moving companies to green up their business activities. Under pressure from the environment and the citizens, the government is getting involved in the production and supply chain of enterprises in various ways. Governments are actively implementing policies to maximize the sustainability of industrial production. They consider social, environmental, and financial issues when making core policies to solve problem [6]. Due to these strategic motivations and pressures from various stakeholders, various companies are adopting green chain supply practices.

15.6 Applications of GSC

Companies are now incorporating environmental initiatives into the various phases of their supply chain: designing, sourcing, manufacturing, and forward and reverse logistics. Just like the traditional supply chain, the GSC practices are used in manufacturing, automotive industry, healthcare, food industry, and pharmaceutical industry.

- *Manufacturing industry:* The manufacturing industry constitutes the pillar industry in the national economy. However, of all business operations, manufacturing is regarded as having the greatest impact on the environment in the form of environmental pollution, generating waste, disrupting the ecosystem, and depleting natural resources. Manufacturing industries are facing pressure from global market to improve their sustainability performance. Green practice adoption is a means of reducing environmental pollution. GSC of manufacturing industry pursues the minimization of negative effect on environment and saves energy. Typically, the price of green product will increase in the manufacturing industry implementing GSC operation [3].
- *Automotive industry:* The automotive industry is one of the largest industries in the world. It brings enormous benefits but also causes troubles to the environment. The pollution from the automotive industry is a problem today. GSCM realizes the pursuit of reducing the pollution and enhances the core competitiveness of the industry. It pursues the minimization of negative effect on environment [7]. The automotive industry may be regarded as a prototype for an implementation of GSC. To meet the emission reduction requirement proposed by the US Congress, some emissions control devices were installed to make the vehicles greener.
- *Food industry:* The production of food plays an important role in agricultural production. The supply chain for food leaves low impact on the environment and requires high resource utilization as well as freshness of food [8].
- *Healthcare industry:* Green and healthy hospital initiatives have improved quality of life for patients and medical staff. For healthcare organizations to reduce environmental hazards that may cause diseases, greening the supply chain is important for their healthcare facilities. Therefore, it is important that these healthcare organizations follow GSC practices [9].
- *Pharmaceutical industry:* The pharmaceutical industry plays a unique role in delivering life-saving products/services to our society. The main objective of the pharmaceutical industry is to build the necessary support systems for healthcare by providing essential medicines. GSC initiatives can help to improve the pharmaceutical industry through recycling unwanted medicines and disposing of medicines in an eco-friendly manner [10].

Other applications include electronics industry, machinery, maritime industry, mining companies, and mobile phone industry. All these other industries are also under increasing pressure to become more sustainable in their supply chains and internal operations.

15.7 Supply Chain Management

The concept of supply chain management (SCM) was developed in the 1970s by focusing on outsourcing, assembling, and delivery of products to customers. The past decade has seen a high demand and development of SCM. The supply chain is the movement of materials as they move from their source to the end customer. Supply chains play a crucial role even in service dominant economies. They are embodied in all sectors such as in automobiles, retail, and IT.

Traditional SCM is only concerned with the delivery of products from buyer to customer. It does not focus on society and the environment. Making green the supply chain is the process of shifting focus from cost or quality issues to also include the environmental and social dimensions of sustainability. The environmental concerns led the companies to incorporate mechanisms regarding disposal, recovery, recycling, and reuse of material. As the supply chain becomes more global, sustainability of supply chain deserves a broader attention [11].

15.8 Green Supply Chain Management

The concept of green supply chain originates from the notion of green purchasing proposed by Webb in 1994. Green supply chain management (GSCM) or sustainable SCM, is a comprehensive philosophy developed recently to support companies and governments to improve their environmental sustainability. GSCM is different from SCM in terms of goal, management structure, business model, business process, and consumption pattern. The evolution of GSCM can be attributed to the increasing awareness among governments, organizations, and customers on increasing pollution, carbon emissions, and deteriorating environmental conditions [12].

GSCM is based on two concepts: the SCM and the environmental management. Thus, the goal of GSCM is to integrate environmental thinking into the traditional SCM. GSCM consists of green purchasing, product design, green logistics, green manufacturing, green packaging, green marketing, and green recycling [13, 14]. These are known as GSCM determinants and are illustrated in Figure 15.3.

- *Green purchasing:* This is a process through which environmentally preferable goods and services are selected. It tries to minimize the environmental impacts of selected products and services.
- *Green design:* It is the process of designing environmentally compatible products and services.

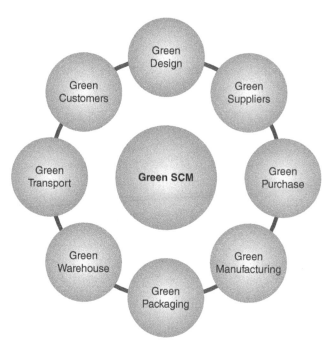

FIGURE 15.3
Different components of GSCM.

Source: http://www.greenhome.com/blog/green-supply-chain-management

- *Reverse logistics:* This involves greening the supply chain backward, that is, starting from the downstream supply chain toward the upstream.
- *Green manufacturing:* This is the renewal of production processes and the establishment of environment-friendly operations within the manufacturing sector.
- *Green packaging:* Packaging design is important for attaining a company's environmental objectives.
- *Green marketing:* This involves the promotion or advertising of products. It includes environment-friendly packaging and distribution. The green philosophy of environmental protection is used as guide throughout the marketing process.
- *Green recycling:* This involves recycling products and packages. It aims at having waste products recycled through manufacturing and remanufacturing.

GSCM is important in many economic sectors such as construction industry, manufacturing, paper industry, automotive industry, maritime industry, power industry, chemical/petroleum industry, and electronics industry.

15.9 Benefits and Challenges

GSCs provide a wide range of benefits. The most immediate benefits are reduced environmental harm and operation costs. Green initiatives can often be cost savers. The GSC brings more economic benefits than the conventional supply chain. It not only improves the environmental performance but also adds value to the business. GSC initiatives are adopted to help reduce costs, increase efficiency, enhance customer satisfaction, increase market share, manage risk more effectively, and increase competitiveness [15]. Deploying a greener supply chain is a win-win scenario for a company, its shareholders, and the planet. GSC will help gain a competitive advantage and attract customers. It is a technical improvement in products and processes intended to enhance resource efficiency. It helps promote lean operations, earn the admiration of customers, enhance loyalty of employees, and improve profitability. It mitigates regulatory burdens and litigation risk [16]. Some of the benefits of GSC are illustrated in Figure 15.4 [13].

GSCM has several advantages. It helps a company to gain a competitive advantage and attract new customers. It improves financial performance and reduces risk by avoiding hazardous material. Other benefits for a company include greater efficiency of assets, improve brand image, less waste elimination, greater innovation, reduction of production costs, mitigation of risks and innovations, reuse of raw materials, improved productivity, increased profitability, and improved overall company performance [17]. Leading multinational corporations such as Walmart and P&G have achieved tremendous results in the practice of SCM.

GSCs face a number of challenges. Large companies tend to resist change because of the large capital and infrastructural investments in the status quo. Although customers may not be willing to pay more for green products, they would prefer them if their prices are the same as the standard ones. Barriers to GSC development include fear of large investments, cost, and the lack of knowledge. The globalization of the supply chain has affected how organizations manage their supply chains. It requires organizations to study cultures, policies, and norms throughout the world for thoughtful and competitive supply chains [18, 19]. There are no regulatory bodies that formulate regulations to meet societal and ecological concerns. The lack of standards could lead to charges of greenwashing.

There are still challenges in the implementation of GSCM. These include significant costs, the complexity of coordination, lack of green training, and lack of communication within the whole supply chain. Market demand is a major external barrier. Lack of top management commitment is a crucial barrier to the successful implementation of GSCM. The implementation of reverse logistics practices in organizations is challenging for both the economy and the ecology [20].

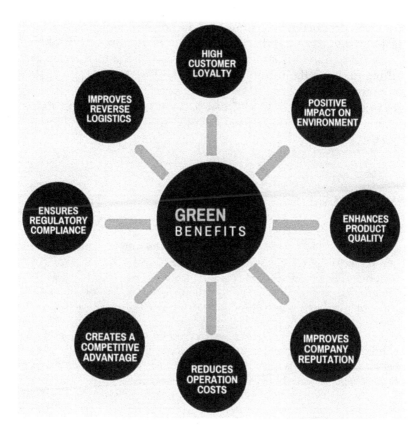

FIGURE 15.4
Benefits of green supply chain [13].

15.10 Conclusion

GSC issues have drawn the considerable attention worldwide of researchers, academics, and practitioners. A GSC requires that suppliers to consider products and environment-related management simultaneously, thereby adding environmental value to products. Robust supply chains have great benefits and should be considered a good investment.

· In this era of globalization, supply chain globalization requires that organizations consider both the cultures and policies of the source companies and their customers' situations. Although governments, societies, and business organizations all over the world support GSC initiatives, GSC is still an evolving concept that will continue to be an important research agenda.

GSCM refers to the concept of integrating sustainable environmental processes into the traditional supply chain. It is integrating environmental thinking into SCM. It is integrated into the entire process including planning, procurement, production, consumption, and reverse logistics. It is an evolving, fast-moving, and multidisciplinary field. Many forward thinking companies are using the environmental issues to their advantage because it is the right thing to do. In the future, companies will be moving to a sustainable supply chain.

More information about GSCM can be found in the book in References [21–29] and the journal exclusively devoted to it: *Logistics & Supply Chain Management*.

References

1. B. Benzamen, "Discrete event simulation of green supply chain with traffic congestion factor," *Proceedings of the 2016 Winter Simulation Conference*, Arlington, Virginia, 2016, pp. 1654–1665.
2. "Improving conventional supply chains," https://saylordotorg.github.io/text_sustainability-innovation-and-entrepreneurship/s10-01-green-supply-chains.html
3. D. Du, F. Zhen, and H. Zhao, "Research on the price negotiation mechanism of green supply chain of manufacturing industry from the angle of customer behavior," *Proceedings of the International Conference on Management Science & Engineering*, Rome, Italy, September 2011, pp. 244–249.
4. P. K. James, "Green supply chain management," January 2013 http://www.infosysblogs.com/supply-chain/2013/01/post_4.htm'l
5. J. Sarki, "A strategic decision framework for green supply chain management," *Journal of Cleaner Production*, vol. 11, 2003, pp. 397–409.
6. J. Gao et al., "Active or passive? Sustainable manufacturing in the direct-channel green supply chain: A perspective of two types of green product designs," *Transportation Research Part D*, vol. 65, 2018, pp. 332–354.
7. Z. Yongan and L. Menghan, "Research on green supply chain design for automotive industry based on green SCOR model," *Proceedings of International Conference on Information Management, Innovation Management and Industrial Engineering*, 2015.
8. Y. Li and Z. Qiao, "Analysis of influencing factors and key driving force concerning the efficiency of green supply chain of fruits and vegetables," *Asian Agricultural Research*, vol. 6, no. 7, 2014, pp. 17–21.
9. S. Balan and S. Conlon, "Text analysis of green supply chain practices in healthcare," *Journal of Computer Information Systems*, vol. 58, no. 1, 2018, pp. 30–38.
10. A. Kumar et al., "When risks need attention: Adoption of green supply chain initiatives in the pharmaceutical industry," *International Journal of Production Research*, 2018.
11. M. N. O. Sadiku, K. G. Eze, and S. M. Musa, "Supply chain management," *International Journal of Engineering Research*, vol. 7, no. 8, August 2018, pp. 137–139.
12. M. N. O. Sadiku, A. A. Omotoso, and S. M. Musa, "Green supply chain management," *International Journal of Trend in Scientific Research and Development*, vol. 3, no. 2, January–February 2019, pp. 901–902.

13. N. Gajendrum, "Green supply chain management—Benefits challenges and other related concepts," *International Journal of Applied Science Engineering & Management*, vol. 3, no. 8, 2017.
14. M. A. Wibowo, N. U. Handayani, and A. Mustikasarn, "Factors for implementing green supply chain management in the construction industry," *Journal of Industrial Engineering and Management*, vol. 11, no. 4, 2018, pp. 651–679.
15. B. Tundys and T. Wisniewski, "The selected method and tools for performance measurement in the green supply chain—Survey analysis in Poland," *Sustainability*, vol. 10, 2018.
16. "Green supply chains," https://saylordotorg.github.io/text_sustainability-innovation-and-entrepreneurship/s10-01-green-supply-chains.html
17. "Bid goodbye to blue and red, color your supply chain 'green' instead," https://www.spendedge.com/blogs/bid-goodbye-blues-red-color-supply-chain-green-instead
18. A. Ali et al., "Green supply chain management—Food for thought?" *International Journal of Logistics Research and Applications*, vol. 20, no. 1, 2017, pp. 22–38.
19. S. A. Al Khattab, A. H. Abu-Rumman, and M. M. Massad, "The impact of the green supply chain management on environmental-based marketing performance," *Journal of Service Science and Management*, vol. 8, 2015, pp. 588–597.
20. V. Balon, A. K. Sharma, and M. K. Barua, "Assessment of barriers in green supply chain management using ISM: A case study of the automobile industry in India," *Global Business Review*, vol. 17, no. 1, 2016, pp. 116–135.
21. J. Morana, *Sustainable Supply Chain Management*. John Wiley & Sons, 2013.
22. S. Emmett and V. Sood, *Green Supply Chains: An Action Manifesto*. John Wiley & Sons, 2010.
23. J. Sarkis and Y. Dou, *Green Supply Chain Management: A Concise Introduction*. Routledge, 2017.
24. J. Sarkis, *Green Supply Chain Management*. New York, NY: ASME, 2014.
25. C. Achillas et al., *Green Supply Chain Management*. Taylor & Francis, 2018.
26. S. S. Ali, R. Kaur, and J. A. M. Saucedo, *Best Practices in Green Supply Chain Management: A Developing Country Perspective*. Emerald Publishing Limited, 2019.
27. A. K. Sahu and A. K. Sahu, *Green Supply Chain Management: An approach towards Sustainability*. LAP LAMBERT Academic Publishing, 2017.
28. T. Paksoy and G. W. Weber, *Lean and Green Supply Chain Management Optimization Models and Algorithms*. Springer International Publishing, 2019.
29. B. Sezen and S. Y. Çankaya, *Ethics and Sustainability in Global Supply Chain Management*. IGI Global, 2017.

16

Green Logistics and Transportation

Vision without action is a daydream. Action without vision is a nightmare.

—Japanese Proverb

16.1 Introduction

We cannot do without transportation in this modern age. It is transportation that allows us to enjoy fresh fruits every day, receive packages overnight, and have Christmas presents delivered to our door just in time [1]. Logistics constitutes the heart of the operation of modern transport systems. The term is now widely used to describe the transport, storage, and handling of products as they move from raw material source to consumption. Modern economy depends on logistics to support the flow of goods. Logistics and transportation are part of the supply chain that deals with the movement and storage of material and products along the supply chain. Logistics and transport activities are well known to have a major impact on the environment. Greening of logistics and transportation operations involve the incorporation of environmental measures. This is necessary in order to achieve the goals of sustainable supply chains [2].

Logistics involves concepts related to the production, distribution, consumption, and disposal. It may be regarded as a tool for moving raw materials, goods, and people to the right place at the desired time. But traditional logistics activities consume a lot of resources and cause much pollution. It cannot meet the requirements of modern society due to huge impact on the environment. To develop modern logistics, environment concerns should be given priority. Green logistics basically consists of all activities related to the eco-efficient management of the forward and reverse flows of products between the point of origin and the point of consumption.

The primary goal of transportation is to mobilize people and goods in an efficient manner from their origin to destination. The transportation sector has made its essential contributions to our lives, from food and transport of goods to personal mobility. The direct impacts of modern transportation activities include personal mobility, economic productivity, and traffic congestion. The transportation sector is energy intensive and has high direct emissions. Green transportation systems are introduced worldwide to reduce carbon emissions. It involves efficient and effective use of resources, modification of the transport infrastructure, and making healthier travel choices. Energy-efficient transportation system helps to achieve the goal of smart and green cities.

This chapter provides an introduction to green logistics and green transportation. It encourages all stakeholders to consider the impact of their actions on the environment. It begins with describing the conventional logistics and reverse logistics. It addresses what greening means to logistics. It highlights the concept and components of green logistics, the parties involved in green logistics, its applications, and how to enhance awareness about it. It addresses benefits and challenges of green logistics. It considers the green transportation, its strategies, benefits, and challenges. The last section concludes the chapter.

16.2 Logistics and Reverse Logistics

Logistics is the term used to describe the transportation, storage, inventory management, order processing, packaging, and handling of products as they move from the source of raw materials, through the production system to their final point of consumption. It focuses on activities such as procurement, distribution, maintenance, and inventory management [3]. The primary purpose of logistics is to reduce costs, especially transport costs. Logistics is a critical factor in promoting globalization and international flows of commerce. For example, transportation is the major activity of most logistics services and it is a logistics operation that has substantial impact on the environment. City logistics aims to achieve the goals of mobility, sustainability, and liveability by balancing the smart growth of economy and cleaner environment. The achievement of these goals ensures the efficient and environmentally friendly urban transport systems.

Reverse logistics is a part of logistics. Reverse logistics and forward logistics are the two subsystems of green logistics. The reverse logistics is a key constituent of green logistics and it is an important component of a business' mission. If logistics involves the movement of material from the point of origin toward the point of consumption, reverse logistics should be the movement of material from the point of consumption toward the point of origin. Reverse logistics activities include remanufacturing, reusing, recycling, landfilling, repackaging, processing returned merchandise due to damage, and salvage. Remanufacturing and rebuilding consume significant amount of resources. Returned products are more difficult and costly to handle than original products [4]. Typical reverse logistics activities would be stock balancing returns and packaging and shipping materials from the end-user. There are eight steps to be successful in reverse logistics [5]:

1. Analyze your returns process and the reasons for returns; then develop detailed returns reporting.
2. Develop a detailed analysis of your current returns-processing costs.
3. Analyze your customers' expectations in terms of speed for processing the returns and issuing credit.
4. Count on resalable returns to fill customer orders and factor returns into your initial buying plans.
5. Determine your requirements, functions, and business metrics.
6. Evaluate all your options for returns processing before selecting a third-party vendor.
7. Compare your business requirements against all viable options, and check references from all vendors.
8. Evaluate all the alternatives and costs in conjunction with product turnaround time and crediting customers.

Like other industries, logistics companies have been dealing with the problems of energy consumption, air pollution, and resource wastefulness. Consequently, the concept of green logistics has emerged. It is a new logistics concept that was proposed by scholars in the West in 1990 [6].

16.3 Green Logistics

Green logistics refers to the systematic measurement, analysis, and mitigation of the environmental impact of logistics activities. It is illustrated in Figure 16.1 [7]. The logistical activities consist of freight transportation, storage, inventory management, materials handling, loading and unloading, packaging, distribution, and related information gathering and processing. The main objective of green logistics is to measure and minimize the ecological impact of logistics activities. Green logistics focuses on material handling, waste management, packaging, and transport. It is closely related to green production, green marketing, and green consumption. There is a strong evidence that green logistics results in increased supply chain performance.

As shown in Figure 16.2 [8], there are strong interactions between logistics, environment, society, and economy. A society without a stable, strong economy will not be able to focus on the environmental or social

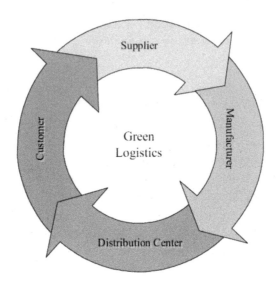

FIGURE 16.1
Green logistics [7].

issues. Green logistics, green manufacturing, and green consumption form a green economic circulatory system, which protects the environment while conserving resources. Green logistics seeks to reduce costs and achieve a more sustainable balance between economic, environmental, and social objectives.

FIGURE 16.2
Interactions between green logistics, environment, society, and economy [8].

The key role of green logistics in the modern economy has been recognized by all levels of government. Government can create a favorable environment for the development of green logistics and take a leading role. Approaches to reducing logistics problems include laws, regulations, taxation, finance, and fiscal subsidies. The US government pays special attention to green logistics. It has formulated several policies and regulations such as pollution source control, traffic volume limit, and traffic flow control [9].

Due to the deterioration of environment and the consumption and diminishing energy sources, green logistics is gaining more and more attention among researchers and industrial practitioners. Green logistics has become an important component of production system in today's world.

16.3.1 Components of Green Logistics

Green logistics has been an integral component of sustainable development of today's economy. The key elements of green logistics include green purchasing, green production, green distribution, green transportation, green storage, green packaging, green loading and unloading, green information gathering and management, green marketing, green consumption, green waste disposal, and green reverse logistics [10–12].

- *Green purchasing:* This is the practice of purchasing goods and services with negative environmental effects in the least possible amounts. It requires various departments within an enterprise to consider environmental factors in the procurement by reducing the costs of material use and cutting costs of end treatment. Manufacturers should consider environmental factors when selecting raw materials.

- *Green production:* This is the implementation of preventative environmental management strategies in a manner integrated in the production process. It involves the idea of producing goods and services with less waste. Green production is reasonable because of the benefits that it bestows on the environment and also because of its fundamental strategic soundness.

- *Green distribution:* This refers to any means of transportation of goods between vendor and consumer with lowest possible impact on the ecological and social environment. It is concerned with the strategies to reduce the environmental impacts of physical distribution. This is a combination of policy measures and pollution-free distribution processing out of consideration for environmental protection.

- *Green transportation:* This is characterized by energy saving, reductions in exhaust emissions, and reduction of air pollution.

Transportation is the most visible aspect of supply chains and is one of the main sources of air pollution. Leakage during transportation should be avoided to prevent environmental pollution. Using green vehicles is also important.

- *Green storage:* This involves creating secure storage environment and keeping products safe in the process of warehouse operations. Storage solutions differ widely in their complexity, configuration, size, and power consumption. Green storage is environmentally friendly, and it incurs lower costs, preserves energy, and improves efficiency.

- *Green packaging:* This is the ecological packaging that can economically meet the functional requirements of packaging throughout its life cycle of product packaging that can be reused or recycled. Green packaging enables packaging of lightweight, recyclable, reuse, recycling, and biodegradable materials. This should cost less and be environmentally friendly.

- *Green disposal:* Waste generation is an inevitable product of human activities. The waste includes common everyday waste such as food waste (e.g., fruits and vegetables), plastics, papers, polythene, metals, batteries, and textiles. Emerging kinds of waste include e-waste, mobile phones, computers, and other kinds of electronic gadgets. Green wastes can be managed through either "grasscycling," or backyard composting [13].

- *Green reverse logistics:* This consists of recycling of unwanted materials (waste materials, boxes, bottles, papers, etc.). It involves remanufacturing, reusing, and recycling. It plays a crucial role in green logistics and green supply chain management. The comparison of reverse logistics and green logistics is made in Figure 16.3 [4].

All these elements impact the environment.

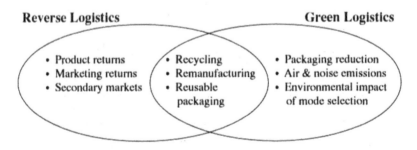

Reverse Logistics   **Green Logistics**

- Product returns   - Recycling   - Packaging reduction
- Marketing returns   - Remanufacturing   - Air & noise emissions
- Secondary markets   - Reusable   - Environmental impact
                             packaging   of mode selection

FIGURE 16.3
Comparison of reverse logistics and green logistics [4].

16.3.2 Parties Involved in Green Logistics

Parties involved with green logistics along the entire supply chain management process include [14]:

- *Government:* The government plays a crucial leading role in the development of green logistics. It is in charge of implementing eco-friendly and environmentally aware policies as wells as national and international regulations. The government should make laws on resource exploitation, using of new materials, and waste recycling. It can promote policies, regulations, and standards to promote the development of green logistics.
- *Customers:* We are all part of the global village and should all make contributions to the environment. The consumers need to understand the type of products they are consuming and demand more eco-friendly ones that involve a green logistics process.
- *Employees:* The employees must ensure that all business operations including logistics are green. They should be willing to work only at environmentally and socially responsible organizations.
- *Society:* The entire society should actively advocate green consumption and supervise enterprises' green logistics. They should be demanding companies to be more aware and change their policies to environmentally friendly ones.
- *Companies:* Companies represent the key players in green logistics development. They should assume responsibility for promoting green logistics. Green logistics should be considered throughout the process in the business. Companies need to find their own motivation and change their policies. They should provide opportunities for their employees to develop green logistics skills.

In addition to these parties, we can also find lenders, insurers, investors, and suppliers.

16.3.3 Applications of Green Logistics

Logistics is an instrument that provides raw materials, products, and people to be in the right place at the desired time. Logistics activities are regarded as one of the basic elements of modern economies around the world. Its practices are adopted in transportation, automotive industry, pharmaceutical sector, smart city, beverage wholesalers, and food distribution. They are also used in efficient uses of electricity, water, and heating in storage facilities.

- *Automotive sector:* In automotive sector, logistics involves the integration of manufacturing, assembly, and distribution activities. Automotive manufacturing involves a large-scale production, transportation, and storage of products. It generates large waste and pollution. Green logistics is an important aspect of modern automotive production system. It focuses on the reduction of waste, environmental hazards, and gas emissions during manufacturing operations [15].

- *Pharmaceutical sector:* Logistics activities play a prominent role in enabling manufacturers, distribution channels, and pharmacies to work in harmony. The need for quick response, recalls, counterfeit products, and the necessity to take some measures to protect efficacy of products make logistics more important in the pharmaceutical sector [16].

- *Smart green city:* Many cities are concerned with the ongoing process of environmental degradation resulting from transportation, manufacturing, and a significant depletion of natural resources. City logistics is intentionally structured and integrated regarding movement of people, materials, and information in an urban area. Green logistics have the effect of practically reducing the negative environmental impacts of households and entities operating in the city [17]. Green city logistics is illustrated in Figure 16.4 [2].

- *Food distribution:* Food is produced and consumed in every part of the world. Food production caused CO_2 emissions. Part of the emissions is the food wasted within supply chains. Based on industrial surveys, logistics costs 10–15% of the price of the food product. This makes food logistics important to producers and retailers [18].

Other areas of application include supply chain management and furniture industry.

16.3.4 Green Logistics Awareness

Sustainability goals can be achieved by rising up people's awareness about their responsibility and rules toward their society. The broad awareness of green concept is the key to long-term development of green logistics. Being a newly emerging concept, people's awareness of green logistics is limited and green logistics professionals are scarce. Logistics professional education lags behind due to lack of logistics teachers. Academic institutions and enterprises should cooperate effectively to provide training, education, research, development, and promotion of new logistics technology. They should prepare more qualified personnel for green logistics industry.

16.3.5 Benefits and Challenges

Today the logistics industry is one of the most important industries by organizing the worldwide supply chains. Green logistics gives prominence to economizing resources and reducing damage to environment. Its aim is to

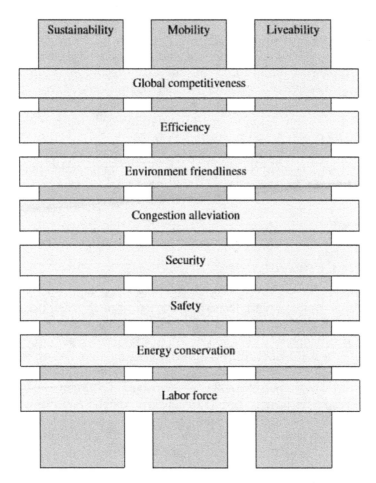

FIGURE 16.4
Green city logistics [2].

reduce logistics costs and increase profits in order to maintain sustainable development. It guarantees better results for the conservation of the environment along the supply chain. It can bring considerable social value and economic value to enterprises and make them obtain a new competitive advantage. It offers solutions to a variety of issues such as consumerism, employee education and training, occupational health and safety, and hunger and homelessness. It can also upgrade enterprises' international competitiveness [19]. Other tangible benefits implied in the application of green logistics include strategic sourcing, overall cost efficiency, improved company's reputation and market share, lower energy consumption, and cost savings caused by reduced fuel and resources.

In spite of the increasing importance of green logistics practices, it is not free from challenges. In order to see how various greening processes can be enhanced, we must also understand the obstacles, that is, what contributes to de-greening. Changing a company's policies and technology to become greener can be expensive. Studies have shown that green logistics implementations add to the costs of investment, operations, and purchasing of eco-friendly materials. There is lack of education and training related to green logistics. Training on new technological processes introduced as part of a green logistics program needs to be implemented. While the least polluting transportation methods such as ships and railways are also the least reliable in terms of on-time delivery and safety, the most reliable transportation modes (such as trucks and planes) are the least eco-friendly ones [20].

16.4 Green Transportation

Transportation is regarded as one of the most important aspects of logistics. It is important to reduce the carbon emission due to transportation. The idea of "green transportation system" was introduced in 1994 by a Canadian named Chris Bradshaw. The objective is to solve transportation problems by developing the traffic tools to decrease traffic jam, to reduce environmental pollution, to boost social fairness, to reduce energy consumption, and to save construction expenses [21]. To solve the air pollution problem and to achieve an environment-friendly transport system, automobile and power generation companies proposed a wide variety of products and solutions.

Green transportation is a highly interdisciplinary area with researchers from different disciplines, including automotive engineers, policy makers, urban planners, and chemical engineers. It seeks to link transportation and environmental concerns. It may be regarded as consisting of four components.

1. *Energy consumption:* Transportation agencies in various countries, along with several standardization organizations, have proposed different types of energy sources (such as hydrogen, biodiesel, electric, and hybrid technologies) as alternatives to fossil fuel to achieve a more eco-friendly and sustainable environment.

2. *Electric vehicles:* These are a key to future clean, green transportation system. A green vehicle (or eco-friendly vehicle) is the vehicle that produces less harmful impact on the environment than conventional vehicles running on gasoline or diesel. Green vehicles can be powered by renewable sources of energy such as wind, solar, biofuels, and hydroelectricity. Electric vehicle charging is a major issue for the massive production of electric vehicles. Electric cars come with a new type of charging station. Major automobile industries worldwide are developing their strategies for wireless charging

technologies. The high costs of electric vehicles, charging stations, and associated infrastructure are challenging.

3. *Smart parking:* Parking is often difficult in a large city. Smart solutions optimize the use of parking lots by equipping each parking space with sensors that detect whether a car is parked there or not. Smart parking in smart cities helps to avoid idle cruising and integrates crowdsourcing with the traditional road navigation system to collect and share real-time information about parking availability.

4. *Society:* The modern society relies heavily on efficient modes of transportation. It is the public thinking and acceptance that always determine the implementation of an idea. Education on green transportation can have significant impacts. Public education campaigns can raise awareness of the benefits of green transportation infrastructure technologies. Governments or private industries have not done much to implement green transportation infrastructure because of technical, regulatory, and social barriers involved. Government policy and commitment are critical in achieving green transportation. However, some lawmakers in California have considered charging green transportation taxes and fees in order to raise additional revenues for transportation [22].

16.4.1 Modes of Green Transportation

Walking, as a mode of transport, is the most sustainable. One should prefer to walk to the school, to work, to grocery shopping, etc. since walking involves zero emission in addition to the health benefits of exercising. Besides walking, there are various modes of green transportation available. Typical average traveling share ratio for the modes of transportation is shown in Figure 16.5 [23]. Typical ones include the following [24].

1. *Bicycle:* Riding a bicycle to work instead of driving a car is great mode of green transportation. People should walk or use bicycles more often.

2. *Electric vehicles:* Common electric vehicles include electric bikes, cars, motorcycles, trains, boats, and scooters. They do not emit any dangerous gasses since they are powered by electricity or renewable technologies like hydroelectric, solar power, and wind turbines. Bus is the dominant motorized vehicle in most big cities. Multiple occupant vehicles (or carpools) reduce the number of vehicles on roads and are favorable mode of green transportation. Car sharing has been going on in the developing world, where traffic and urban density is worse. Green vehicles are more fuel efficient and should have less environmental impact than equivalent standard vehicles.

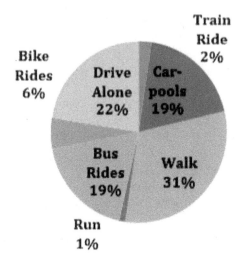

FIGURE 16.5
Typical average traveling share ratio [23].

Sustainability in Internet of vehicles can be achieved by the use of pollution-free vehicular systems and by maintaining road traffic safety or prevention of accidents.

3. *Green trains:* Trains are getting increasingly greener with innovative green technologies. These electric trains travel at tops speeds of more than 200 mph.

16.4.2 Strategies on Green Transportation

These strategies are all based on the principle of sustainable development [25].

1. Call for using public transport vehicles instead of private cars.
2. Improve new energy vehicles and give priority to rail, electric, or hybrid vehicles.
3. Construct walking and bicycle systems.
4. Develop intelligent transport system.
5. Strengthen enforcement on green transport and education systems.
6. Make efforts, the government should, to establish an efficient public transportation system.

16.4.3 Benefits and Challenges

There are several benefits of green transportation that will enhance healthier lifestyle, improve quality of human life, and allow sustainable mobility. Green modes of transportation have produced zero emissions. They play an important role in achieving a green city. Green transportation can also enhance quality of life and certain economic activity.

The materials used in manufacturing electric batteries are scarce and costly. There is lack of global standards and regulations for electric vehicles. Home charging capacities are limited to one or two cars at most [26].

16.5 Conclusion

Logistics is an important part of the modern economies around the world. Green logistics is an emerging cutting-edge concept that ensures that environmental, social, and financial-economic factors are taken into account in the production and distribution of products in the market. It is one of the main factors for sustainable development. It provides a useful tool for eco-logistic and sustainable enterprises. Its adoption may save costs and reduce environmental damage while improving the operating efficiency of logistics enterprises.

Transportation is one of the most indispensable means of life. It plays a major role in people's everyday lives and is a decisive factor in economic competitiveness. Green transportation is a low-carbon and environment-friendly mode. Green transportation modes include walking, bicycle, public transport, and rail transport [27]. Many cities worldwide have realized the need to link sustainability and transport policies. Green transport is now becoming a strategic choice for most countries.

For more information on green logistics and green transportation, one should consult the books in References [2, 28–34] and the journals devoted to them:

- *International Journal of Logistics*
- *International Journal of Logistics Research and Application*
- *Journal of Business Logistics*
- *Contemporary Logistics*
- *International Journal of Sustainable Transportation*
- *Journal of Advanced Transportation*
- *Transportation Research Part D*

References

1. P. J. Katsioloudis and M. V. Jones, "Green transportation for a green earth," *Technology and Engineering Teacher*, April 2012, pp. 19–25.

2. B. Fahimnia et al., *Green Logistics and Transportation: A Sustainable Supply Chain Perspective.* Springer, 2015.

3. D. Gechevski et al., "Reverse logistics and green logistics way to improving the environmental sustainability," *Acta Technica Corviniensis—Bulletin of Engineering,* January–March 2016, pp. 63–70.

4. D. S. Rogers and R. Tibben-Lembke, "An examination of reverse logistics practices," *Journal of Business Logistics,* vol. 22, no. 2, 2001, pp. 129–148.

5. S. Nylund, "Reverse logistics and green logistics: A comparison between Wärtsilä and IKEA," https://www.theseus.fi/bitstream/handle/10024/46993/Reverse%20Logistics%20and%20green%20logistics.pdf

6. S. C. Tang et al., "Strategies for the development of green logistics in Taiwan," *Proceedings of Economic and Social Development,* April 2013, pp. 1335–1345.

7. S. Zhang, "Swarm intelligence applied in green logistics: A literature review," *Engineering Applications of Artificial Intelligence,* vol. 37, 2015, pp. 154–169.

8. "Green logistics," *Wikipedia,* the free encyclopedia, https://en.wikipedia.org/wiki/Green_logistics

9. L. Xue, "Countermeasure research of green logistics development in Shandong province," *Proceedings of the 15th International Conference on Control, Automation and Systems,* Busan, Korea, October 2015, pp. 1672–1675.

10. Z. Guirong and M. Yaxin, "Green logistics management of logistics enterprises," *Proceedings of the 3rd International Conference on Information Management, Innovation Management and Industrial Engineering,* 2010, pp. 567–569.

11. S. T. Rad and Y. S. Gulmez, "Green logistics for sustainability," *International Journal of Management Economics and Business,* vol. 13, no. 3, 2017, pp. 603–614.

12. L. Li and L. Yanmin, "Development path of green logistics under environmental maintenance," *Proceedings of IEEE 3rd International Conference on Communication Software and Networks,* May 2011, pp. 51–57.

13. M. Mohammed, I. Ozbay, and E. Durmusoglu, "Bio-drying waste with high moisture content," *Process Safety and Environmental Protection,* vol. 111, 2017, pp. 420–427.

14. "Green logistics, challenge for SMEs, supplier performance and demand planning – what will and should happen in supply chain management," https://www.ltdmgmt.com/green_logistics.php

15. D. Chhabra, S. K. Garg, and R. K. Singh, "Analyzing alternatives for green logistics in an Indian automotive," *Journal of Cleaner Production,* vol. 167, 2017, pp. 962–969.

16. M. Arslan and S. Şar, "Examination of environmentally friendly 'green' logistics behavior of managers in the pharmaceutical sector using the theory of planned behavior," *Research in Social and Administrative Pharmacy,* vol. 14, 2018, pp. 1007–1014.

17. M. Jedliński, "The position of green logistics in sustainable development of a smart green city," *Procedia - Social and Behavioral Sciences,* vol. 151, 2014, pp. 102–111.

18. P. Helo and H. Ala-Harja, "Green logistics in food distribution—A case study," *International Journal of Logistics Research and Applications,* vol. 21, no. 4, 2018, pp. 464–479.

19. C. Rong, "Green logistics research based on sustainable development," *Proceedings of the 2nd International Conference on Artificial Intelligence, Management Science and Electronic Commerce,* August 2011, pp. 4635–4639.

20. M. N. O. Sadiku, A. E. Shadare, and S. M. Musa, "Green transportation," *International Journal of Trend in Research and Development*, vol. 6, no. 1, January–February 2019, pp. 213–214.

21. D. Zhang and A. Fei, "Green transportation: The essential way for transportation in the future," *Applied Mechanics and Materials*, vols. 97–98, 2011, pp. 1135–1140.

22. A. Weinstein, J. Dill, and H. Nixon, "Green transportation taxes and fees: A survey of public preferences in California," *Transportation Research Part D*, vol. 15, 2010, pp. 189–196.

23. J. Froehlic et al., "UbiGreen: Investigating a mobile tool for tracking and supporting green transportation habits," *Proceedings of the SIGHI Conference on Human Factors in Computing Systems*, Boston, MA, April 2009, pp. 1043–1052.

24. "What is green transportation?" https://www.conserve-energy-future.com/modes-and-benefits-of-green-transportation.php

25. Z. Liu et al., "Green transport practice in Beijing," *The Fifth Advanced Forum on Transportation of China*, Beijing, China, October, 2009, pp. 80–84.

26. S. Mehar et al., "Sustainable transportation management system for a fleet of electric vehicles," *IEEE Transactions on Intelligent Transportation Systems*, vol. 16, no. 3, June, 2015, pp. 1401–1414.

27. H. Li, "Study on green transportation system of international metropolises," *Procedia Engineering*, vol. 137, 2016, pp. 762–771.

28. A. McKinnon, M. Browne, and A. Whiteing (eds.), *Green Logistics: Improving the Environmental Sustainability of Logistics*, 3rd ed. Philadelphia, PA: Kogan Page Publishers, 2015.

29. A. M. Brewer, K. J. Button, and D. A. Hensher (eds.), *Handbook of Logistics and Supply-Chain Management*, vol. 2. Emerald Group Publishing Limited, 2001.

30. R. R. Singh and D. R. Gaur, *Sustainable Logistics and Transportation*. Springer International Publishing, 2017.

31. D. B. Grant, C. Y. Wong, and A. Trautrims, *Sustainable Logistics and Supply Chain Management: Principles and Practices for Sustainable Operations and Management*. Kogan Page, 2017.

32. M. Kutz (ed.), *Environmentally Conscious Transportation*. Hoboken, NJ: John Wiley & Sons, 2008.

33. H. N. Psaraftis (ed.), *Green Transportation Logistics: The Quest for Win-Win Solutions*. Springer, 2015.

34. K. Furgang and A. Furgang, *On the Move: Green Transportation*. Rosen Publishing Group, 2009.

Index